T-Labs Series in Telecommunication Services

More information about this series at http://www.springer.com/series/10013

Maija Elina Poikela

Perceived Privacy in Location-Based Mobile System

 Springer

Maija Elina Poikela
Quality and Usability Lab
Technische Universität Berlin
Berlin, Germany

Zugl.: Berlin, Technische Universität, Diss., 2019

ISSN 2192-2810   ISSN 2192-2829   (electronic)
T-Labs Series in Telecommunication Services
ISBN 978-3-030-34173-2   ISBN 978-3-030-34171-8   (eBook)
https://doi.org/10.1007/978-3-030-34171-8

This Springer imprint is published by the registered company Springer Nature Switzerland AG.
The registered company address is: Gewerbestrasse 11, 6330 Cham, Switzerland

Preface

During the past decade, location-based services have become increasingly popular. These services use the physical location of the user to provide various functionalities, ranging from targeted recommendations to social benefits. Such benefits might include arranging meetings or posting visited locations on social networking sites for a favourable self-presentation. Many location-based services are seemingly free for the user. In the business model that is prevalent in today's services, the service providers create their profits from advertising. In order to target the adverts more precisely, and to therefore improve the possibility that the user would follow the advert and thus create monetary benefit for the service provider, these adverts are often based on the user's behaviour, such as browsing history, social media *likes*, or visited locations. These pieces of location information, especially when collected repeatedly or even continuously, can reveal a great deal about the user, including home and work address, hobbies, favourite restaurants, and even medical visits. As a consequence, in order to use the benefits of such a service, the user accepts the risk to one's privacy.

In order to adequately assess whether or not to accept the exchange of private information, the user must understand the privacy risks and evaluate whether they outweigh the expected benefits. In practice users might not make such well-balanced decisions due to lack of proper understanding of data flows in location-based services, because of optimistic expectations that risks do not affect oneself, or for other reasons. This work aims at understanding which factors influence privacy behaviours in the context of location privacy. The aim of this work is to create a model assessing the impact of various factors on location privacy behaviours and finally, based on these factors, to predict privacy behaviour in this context.

Empirical studies are conducted to assess how location privacy behaviours could be predicted and whether or not the users could be positively influenced towards more privacy-conscious behaviours. The findings from this work show that some of the most important factors driving users' privacy behaviour are the influence of one's peers, trust in the data recipient, the assessment of perceived risks and benefits, as well as the payment provided for shared location information.

Different methodological approaches, including quantitative and qualitative methods, are used in the conducted studies throughout the work. These approaches are presented in detail, their benefits and drawbacks are discussed, and finally, methodological recommendations are given for future research.

Berlin, Germany   Maija Elina Poikela

Acknowledgements

During the course of the years that I have been working on this thesis, many individuals have been there in different ways supporting me, and I can only express my most sincere gratitude to all of them.

While working at the Quality and Usability Lab, I have crossed paths with several colleagues who made my stay worthwhile, many of whom have become friends to me. There were also many individuals who gave invaluable support in the scientific work. Most importantly, I would like to thank Ina Wechsung, whose guidance to various topics of statistics made it possible for me to do this Ph.D. I would not be here without her support. I would like to thank the whole Crowdee team for their assistance, and in particular Babak Naderi for all the help he provided with crowdsourcing as well as SEM related matters. I am grateful to many colleagues for the discussions, scientific and otherwise; I would like to especially thank Patrick Ehrenbrink for the moral support during the thesis writing. My thanks go also to Steven Schmidt for translating the abstract into German. I would also like to thank in particular Tobias Hirsch, Irene Hube-Achter, and Yasmin Hillebrenner for their support throughout my stay at the QU Lab. My many thanks go also to Lydia Kraus, Alexander Bajic, Niklas Kirschnick, Hanul Sieger, Klaus-Peter Engelbrecht, Marie-Neige Garcia, Tim Polzehl, Tilo Westermann, Benjamin Bähr, Jan-Niklas Voigt-Antons, Marc Halbrügge, Laura Fernández Gallardo, Benjamin Weiss, Stefan Hillmann, Saman Zadtootaghaj, Rafael Zequeira, Neslihan Iskender, Tanja Kojic, Thilo Michael, Gabriel Mittag, Falk Schiffner, and Carola Trahms, as well as all others who I have had the pleasure to spend time while working at the Q&U Lab. I would also like to thank the students and student workers who supported me in many of the studies during this work, especially Maurice Baiers, Felix Kaiser, Robert Schmidt, and Steven Winter. I am grateful to my advisor Sebastian Möller for providing me support and guidance throughout my PhD, as well as to Eran Toch for his invaluable scientific advice. I also greatly appreciate Marian Margraf for agreeing to be a co-examiner of my dissertation.

I would like to thank all of my family and friends for their endless support and love. This is to each of you: Thank you! The biggest influencers in my professional life have been my dear parents: I can thank my mother for the inspiration for open-mindedness and curiosity, and my father for choosing a career in science and technology. The quality of this text was greatly improved by my brother Antti, who meticulously spotted all spelling mistakes; any remaining ones I have personally added despite his advice. He, together with my other siblings, has over the years shaped my values and thinking—I would like to express my gratitude also to Mikko, Anna, and Kaisa for being you, and making me who I am.

Most importantly, I would like to thank my amazing other half, my husband Rahul, for his support, scientific and otherwise; most of the improvement in my writing skills is because of your guidance. For so many reasons, majority of which are non-scientific, I feel a great privilege to have you by my side. Finally Hilla, the concentrated package of joy, thank you for reminding me to stick to what is essential.

Contents

Acronyms

ACCU	Information accuracy
AVE	Average variance extracted
BENE	Perceived benefits
CCTV	Closed-circuit television
CFA	Confirmatory factor analysis
CFI	Comparative fit index
CFIP	Concern for information privacy
COLL	Data collection
CTRL	Perceived control
DFP	Desire for privacy
EFA	Exploratory factor analysis
GDPR	General data privacy regulation
GPS	Global positioning system
IUICP	Internet users' information privacy concern
KMO	Kaiser-meyer-olkin
KNOW	Location privacy knowledge
LBS	Location-based services
LPBA	Location privacy beliefs and attitudes
LPT	Location privacy taxonomy
LPV	Location privacy valuation
M	Mean
NORM	Perceived norm
PCAL	Privacy calculus
PET	Privacy-enhancing technology
PUSU	Purpose and secondary use
RAA	Reasoned action approach
RISK	Perceived risks
SD	Standard deviation
SNS	Social networking service
TAMP	Technology acceptance model
TPB	Theory of planned behaviour

TRA	Theory of reasoned action
UNAUT	Unauthorized access
UTAUT	Unified theory of acceptance and use of technology
WILL	Willingness to disclose
WTA	Willingness to accept
WTP	Willingness to pay

Chapter 1
Introduction

1.1 Motivation

Smartphones, smartwatches, and other smart gadgets have become widely popular during the past decade. These devices are carried most of the time on or very close to the user's body such as in a pocket or a handbag, making them very personal devices. These devices use various sensors to gather data that is used in different functionalities provided for the user, including sensors measuring the motion, environmental properties such as humidity and brightness, and location of the device. Using these sensors, a huge amount of details of the device—and its user—can be determined [14, 171]. These details include how the device is being held, if the user is typing using the thumb, or if it is carried in a pocket or in a purse [106]. Arguably the most sensitive type of data, however, is the location of the device.

In 2006, Clive Humby stated that data is the new oil—it is useless when unrefined, but highly valuable when properly treated [122]. With today's computing power and storing capacities, refining this new oil has become cheaper than ever. Though, unlike oil, data is not a limited resource, but ever-more data is produced and analysed to extract user behaviour: shopping habits, medical history, and places visited, just to mention but a few. Furthermore, it can be argued that if data is the new oil, then *privacy is the new climate change*.

A vast body of literature since the beginning of the millennium examines users' privacy concerns and preferences in the context of LBS. The aim of this dissertation is primarily to assess how different aspects shape the users' privacy perceptions and attitudes in the context of location privacy, and how these perceptions influence behavioural choices. To limit the scope of the work to a practical but very relevant type of service, this work concentrates on location sharing on smartphones. Nearly 80% of the US adult population [129] owned a smartphone in 2017, and slightly more in the adult German population [154]. Thus, it can be stated that smartphones

© Springer Nature Switzerland AG 2020
M. E. Poikela, *Perceived Privacy in Location-Based Mobile System*,
T-Labs Series in Telecommunication Services,
https://doi.org/10.1007/978-3-030-34171-8_1

are a type of personal device that are so widely adopted that usage patterns with them can be considered to represent a sizeable proportion within the whole general population.

1.2   What Is Privacy?

Privacy is a complex concept that has proved to be difficult to define. Often privacy is considered a human right integral to society's moral value system. Lack of privacy can lead to self-censorship, as individuals restrict their expressions and actions. Free expression can be considered a fundamental right in a free democracy, and therefore, it can be stated that privacy is crucial for society. Privacy is explicitly protected in the European Convention on Human Rights as "the right to respect for private and family life" [53]. In Europe privacy laws protect individuals from humiliation by protecting their self-control over the use of their personal information [175]. Also, the German constitution starts with a phrase stating that "Human dignity shall be inviolable" [45]. In the USA, the main privacy protection is established in the Fourth Amendment as the right against unlawful searches and seizures, mostly within one's home [45]. Thus, cultural differences can be seen in what is understood as privacy, even between countries representing modern western cultures. One of the more accepted views considers privacy as "the selective control of access to the self" [12], a definition which is adopted also in this work.

1.2.1   Perceived Privacy

When considering *perceived privacy*, the focus moves from the absolute control over one's personal information to the perception of the extent to which one is in control of the access to the self. In particular in today's world where data about the users' every action is recorded through various sensors, the user has limited capability of knowing and understanding the extent to which their data has been collected, which instances have access to that data, and what kind of inferences can be drawn from the collected data. It is therefore the users' own perception of their privacy that is likely to dictate their behaviour with respect to different technologies.

More privacy is not necessarily better, but experiencing too much privacy might lead to feelings of seclusion and loneliness, making privacy non-monotonic [126]. Thus, to achieve a desired level of privacy, a person may, depending on the context, use different approaches in their privacy behaviour.

The communication privacy management theory by Petronio [128] states that individuals have control over how and with whom to share their private information by setting boundary rules. According to the theory, after disclosure, the recipient of information becomes a co-owner of the information. If the co-owners fail to effectively agree on the privacy rules, or if the rules are disregarded, the so-called *boundary turbulence*—privacy concern and discomfort—ensues. Such is the situation when the user does not have full control of disclosure, or if their expectations of the subsequent data use are not met.

1.2.2 Information Privacy

With the popularity of smart devices, there are also a growing number of apps created for the users of such devices. In the first quarter of 2018, there were 3.8 million apps available for users to download in the Android Play Store, and two million in the Apple's App Store [155]. For example, for the Android platform, it is a rather straightforward process for anyone with the required knowledge to create and publish an app. As a consequence, also the number of organizations, companies, and other parties potentially collecting personal information from the users increases. It is also typical to forget which apps one has on a device [10], and if some data from the user is collected as a background process, the user can be rather unaware of the data collection and its recipients [164]. Additionally, data collected by an app might also be shared with third parties such as advertisers or other applications, increasing the list of parties accessing the user's data even further; though, according to the European Union's (EU) General Data Privacy Regulation (GDPR), all data collection should have an explicit user consent, and therefore such secondary use of data is, in fact, illegal [160].

While the smart devices have become increasingly popular and the number of applications developed for such devices has skyrocketed, there has also been a radical growth in the amount of mobile malware: over 600,000 new pieces of malware were developed targeting solely Android in just the month of June in 2016 [15]. Since then, the situation has likely not improved.

The concept of privacy often refers to information privacy, and the control that an individual has over how their personal information is collected and used [174]. This control can be seen diminished with the advent of the information age as the users' information has become easily collected, stored, and utilized [149], and users' privacy concerns have been on the rise [1].

There are various new aspects that affect the user's privacy when moving to online context:

- Audiences of shared information can be big and distant both spatially and temporally. A message posted on an online forum today can be shared over and over again, reaching much larger audiences than initially intended. The post might also be still accessible in a couple of decades' time—even to audiences not yet born today.
- Control over how disclosed information is used is diminished. Even though the EU now provides its citizens a "right to be forgotten", removing information that has been shared once is a complicated process. Also, most of the time data

collection happens unbeknownst to the user, but one would need to know that information of oneself is out there in order to protest against such collection.

- Interpretation of the information can change in different contexts. The post in the earlier example could be interpreted in a context that was not intended by the user, giving a new, unintended possible interpretation to the message. To adequately assess whether or not to share the original message, the user should be able to know how their message would be available and interpretable to others also in new contexts.

- Interpretation of the information can change at different times with changing culture, knowledge, politics, and societal norms; as a consequence, the same message might get a new interpretation when other factors change over time, and no longer represent the user's opinion.

Today, communications and transactions are to a great extent taking place in an online environment, where user's every action—every online search, mouse click, and "like"—is potentially tracked, logged, and analysed. Such a situation can be considered to be similar to surveillance. Seen as a symbol for the power hierarchy and surveillance in the society, Michel Foucault referred metaphorically to the *Panopticon* [26], where inmates of a prison can at no point in time be certain whether or not they are being watched [60]. Foucault pointed out that similarly to the Panopticon structure changing the inmates' behaviour towards social norms, being under continuous surveillance through closed-circuit television (CCTV) cameras also creates a pressure to conform to the norms. The same effect, known as the *observer effect* [107], also manifests itself in the online world—users are not sure who might see which links they click, which sites they browse, or which search words they use. When users hesitate to express themselves, freedom of speech can be considered to be at risk because of the self-censorship.

One view to privacy considers it a commodity, stating that privacy has an economical value [2]. Indeed, privacy is an interesting topic for the industry, not only because of the privacy legislation that companies are required to abide to, but also because privacy perceptions influence how users adopt services, or abandon them [183]. According to such a view, users advisedly exchange their personal information to services. Many online services that are seemingly free for the users are financed through advertising. For maximising the income, the adverts need to be well targeted for the user. Therefore much of the online adverts are based on the users' behaviour, such as visited sites or physical locations. This creates discomfort in the users, most of whom find targeted advertising invasive [105]. In particular, location-based advertisements induce concern in mobile users [88]. Also, the adverts based on visited locations can be shown on various devices in the user's network, potentially compromising the user's privacy among other users of the devices in the network.

1.2.3 Location Privacy

For defining location privacy for the purposes of this research, this work extends the definition of privacy to *Selective control of access to one's location or location history*. Location privacy has been defined by Beresford and Stajano correspondingly as "the ability to prevent other parties from learning one's current or past location" [28]. This definition specifies that the data subject—or the person whose location data is affected—should be in control of the use and subsequent use of their data, which can be considered an extension of the earlier presented definition of privacy.

The physical location of the device can be calculated using three main methods:

1. satellite navigation systems such as the Global Positioning System (GPS),
2. using the mobile carrier antennas, and
3. wireless networks.

A device can use any combination of these three methods to calculate the most accurate location of the device. A combination of the methods make the positioning more accurate, but on the other hand, might consume more power, and thus affect the battery life of the device [182].

When the device to be located is outdoors, satellite navigation systems, which provide a method for autonomous geographical positioning, provide positions to be determined with an accuracy of just a few centimetres when several satellites are available [54]. For open access public use, the accuracy is currently at around one meter radius [41]. The accuracy decreases when the signal is blocked by objects such as mountains or buildings, and drops quite drastically for indoor environments [41].

When satellite signals are not available—for example, when the user has switched the GPS off on their device—the geographical position of the device can still be determined using other methods. When the device is on and emitting a signal to the carrier antennas, the position of the device can be estimated using methods based on the strength of the emitted signal, and the geographical location of the antennas. The method, known as *trilateration*, requires at least three measurement points, or antennas, to define the location of the device unanimously [112]. The accuracy of the positioning thus depends on the resolution of the antennas. In urban areas where the population, and the resolution of antennas are denser, precision of around 50 m can be achieved [96].

In indoor environments, satellite based positioning is inadequate to locate a device accurately because of the walls obstructing the signal. Especially in such indoor situations, defining the position of a device can be done using Wi-Fi access points. Estimation of the geographical location of the device happens typically from the MAC addresses of the access points, and using trilateration techniques to calculate the relative position of the device [171]. The accuracy of the method is dependent on the number of access points. Wi-Fi access point based positioning is used also in outdoor environments [96]—and also other methods for indoor positioning exist [171].

The location coordinates of a device can then be conveyed to different services, which use it in order to provide various functionalities for the user. Such services using the spatial location of the device are called *location-based services* (LBS) [96].

The physical location of the device is used in many smartphone applications, or *apps*, such as to inform the user about an optimal route, to connect with other users nearby, and to provide games with more context of the surroundings. The user can share their location actively for instance through social media applications, or, when the user has once given their consent for data collection, the location data can be collected automatically whenever the smartphone is on.

Logged behavioural data can be considered a more accurate source of information than user-filled profiles, as behaviour can be collected from user actions and digital footprints. Converting the physical footprints into such digital ones, location information can be considered one of the most sensitive types of data, as from seemingly anonymous location traces users can be uniquely identified [109]. Unique positioning of a device of a user is possible within a database of 1.5 million users on an accuracy of 95%, even when measured on a resolution of carrier antennas [109]. Thus, even datasets that seem anonymous or blurred provide rather low anonymity. Through building a profile based on these location traces, information, such as home location, not-at-home times, favourite types of restaurants, typical mode of transport, and hobbies can be identified. With enough data, connecting this information to an individual is rather straightforward.

Even a few location samples make it possible to uniquely identify a user and create accurate profiles of them [61], leading to a serious erosion of privacy. Users can be also uniquely identified if their home and work locations can be deduced from the data [63], which can be done in many cases by comparing the location information with the time of the day. This signifies that even if shared location data is anonymous per se, the users' anonymity is not guaranteed. With the complexity of the data collection scenario and the lacking transparency, the user is to a large extent left with no other option than to either trust the service providers not to infringe their privacy or to discontinue using the services. Thus, the seemingly free services are paid for with location data, which can reveal the user's daily habits, and can be rather accurately de-anonymized even from big datasets [109].

There are also other types of LBSs than smartphone applications. As one example, public transport cards could be considered to be LBS. When a user of a public transportation system pays for their travel using a chip card, information about the beginning time and location of the trip is logged, and in some systems, also the end time and location. As another example, radio-frequency identification (RFID) is a technology to automatically track small tags attached to items. The RFID technology is used widely in various use cases, such as to track products throughout the production line, to mark pets, or to tag products in retail [69]. The tags can be unperceivably small so that the consumer might not be aware of its existence when buying a product. The tags remain functional even after the product has been purchased, and because they can be read from a distance, it is possible to gather data from the individual from a distance [69, 159].

It is not only malicious attacks that cause the users privacy concern. Many smartphone applications leave it for the user to define their privacy settings; however, if at all, they give only coarse options for location privacy settings. For example, the current privacy setting options for Android smartphones allow the user to define whether or not to give an application access to the device's location as a binary allow-or-deny decision. If the user does not allow the physical location to be used, many functionalities are not accessible. As a consequence of the lack of options, users may feel that they have lost control over their privacy [98].

The benefit a user receives from using LBS could be an improved service, enhancing one's social status, or enhanced communication [178]. The gains come with a price of a number of risks related to location sharing, including surveillance, inappropriate secondary use of data, or other privacy violations [163, 178]. Therefore, sharing one's location using LBS can be considered a transaction, where a user receives a benefit with the cost of a privacy risk. Being able to understand what happens with the location information after it has been shared with a service provider is a core precondition to adequately assess the risk–benefit ratio of such a transaction; however, collecting personal information might happen without the users' knowledge. While such a collection might be completely according to laws and privacy policies the user has accepted, if the users' expectations of what happens to personal information are not met, the user feels a lack of control over one's privacy [116]. If location privacy is defined as the selective control of access to one's location information, it can be stated that when such information is used without the user's explicit consent, their location privacy has been compromised.

1.3 Research Questions

-This work aims at explaining privacy behaviour in the context of LBSs. There are three main aspects that the work focuses on: the background factors to location privacy, beliefs and attitudes arising from the background factors, and behavioural outcomes, which are postulated to be directly influenced by the attitudes. There are a number of research questions that are addressed within this dissertation; these questions are presented in the following sections.

1.3.1 Background Factors Influencing Location Privacy Beliefs and Attitudes

There are various factors that are in this work assumed to be influencing privacy beliefs and attitudes. More such factors probably exist than can be adequately assessed within this dissertation; here the focus is mainly on user-related factors. The first such factor is privacy concern that, somewhat like a personality factor, is in this work considered to be a rather unchanged trait.

RQ 1.1. Privacy Concern: How does the users' privacy concern influence the privacy beliefs and attitudes in the context of location privacy?

Findings from previous studies state that experiencing privacy violations online has an impact on users' privacy attitudes [150], and behaviour [43]. The second research question tackles the issue of whether there is such an impact also in the context of location privacy.

RQ 1.2. Prior Privacy Violations: How do privacy violations experienced in the past influence privacy beliefs and attitudes in the context of location?

Existing privacy literature suggests that trust can mitigate perceived risks in an online context [48, 92], as well as in the context of LBS [183]. Thus, this work considers trust as a background factor influencing location privacy beliefs and attitudes, and the third research question can be formulated as follows:

RQ 1.3. Trust: What is the influence of trust on location privacy beliefs and attitudes?

Understanding information flows on the services that are used is an important factor in evaluating the state of one's information privacy. Knowledge might, coupled with trust, influence perceived risks in the usage of a service [120], as well as the perception of control. Therefore, the third research question addresses location privacy knowledge as a background factor influencing privacy beliefs and attitudes.

RQ 1.4. LBS Knowledge: How does the users' general knowledge related to data privacy in location-based services affect the privacy beliefs and attitudes?

Previous works have found context to play an important role in decisions concerning location privacy [166]. However, this work does not consider physical context in detail. Social context is addressed in terms of the recipient of information:

RQ 1.5. Social Context: How does the recipient of location information influence location privacy attitudes, and accuracy of location sharing?

1.3.2   Location Privacy Beliefs and Attitudes Influencing Behavioural Outcomes

In this work, behavioural intention as well as privacy behaviour is postulated to be directly influenced by various beliefs and attitudes related to that behaviour. First, the user of LBS evaluates the benefits, as well as the risks, related to usage of such services:

RQ 2.1. Benefits and Risks: What are the benefits and risks that users of LBSs perceive in the usage of such services?

When the benefits as well as risks have been assessed, the two are combined to determine whether the risks or benefits weigh more [47]. Through the risk–benefit assessment, or *privacy calculus*, two questions arise:

RQ 2.2. Privacy Calculus Constituents: How are the perceived benefits and risks weighed?

RQ 2.3. Privacy Calculus: How does the privacy calculus influence behavioural intention, location sharing behaviour, or the usage of privacy-enhancing technologies?

Peers might have an impact on the user's views of how desirable certain behaviour is. In the context of privacy, the peer influence could modify the users' behaviour in terms of disclosing information about oneself: when others disclose personal information freely, such behaviour might be perceived as more desirable, and possibly also safe. Peers have an impact also on usage of privacy protection methods [165].

RQ 2.4. Social Norm: How does the social norm influence privacy behaviour in terms of location disclosure, usage of privacy protection mechanisms, or behavioural intention?

In the context of privacy, perceived control can be defined as the control that a user has over disclosing personal information [11]. Additionally, perceived control could refer to the feeling that there are methods or technical measures at the users' disposal, and that the user has the required knowledge or support in order to use them. This leads to the next research question:

RQ 2.5. Perceived Control: How does the feeling of being in control influence the behavioural intention, or privacy behaviour?

Evaluating risks and benefits is a tedious process, and the result of the assessment is not easily quantifiable. In an attempt to address the issue, a method of monetary quantification is used as a mediator between location privacy beliefs and attitudes, and behavioural outcomes.

This leads to the final research question in this category:

RQ 2.6. Location Privacy Valuation: How can the value of location privacy be quantified?

As location privacy valuation is essentially expected to quantify privacy calculus in the context of location privacy, the research question includes also the issue of whether or not the measured location privacy valuation does indeed relate to privacy calculus to an extent that location privacy valuation could be considered a quantification of privacy calculus. Furthermore, the method is only useful if it has predictive power when it comes to assessing privacy behaviour; thus it becomes necessary to evaluate whether or not privacy behaviour can be predicted from the value the users give to their location information.

1.3.3   Behavioural Outcomes

The final category being addressed in this work addresses various types of behavioural outcomes in the context of location privacy. These behaviours include:

- Intention to disclose location information by using LBS
- Location disclosing behaviour, and
- Location privacy protecting behaviour.

How to predict these behavioural outcomes has largely been addressed within the research questions in the previous category. Next, the relationship of intention and behaviour is addressed:

RQ 3.1. Intention: How does behavioural intention influence behaviour in terms of location disclosing behaviour?

When an individual becomes aware of their behaviour, this awareness might have an influence on the attitudes corresponding to that behaviour. The research question is addressed through *privacy nudges*, which inform the users about their location disclosures. A more practical question on the same topic is whether or not the users' behaviour can be positively influenced towards more conscious privacy decisions by informing them through privacy nudges. This kind of a feedback loop is addressed as follows:

RQ 3.2. Awareness: How does awareness of one's location disclosures influence the privacy beliefs and attitudes?

Finally, the last research question brings the discussed factors together.

RQ 3.3. Behavioural Model: Using the dimensions identified in this work, to what extent can location sharing and protection behaviour be modelled?

1.4   Structure of the Book

This book is organized as follows:
Chapter 2 begins with an overview of existing theories and models of behaviour, as well as privacy behaviour. Then, relying on some of the theories, a taxonomy summarizing the concept of location privacy in the framework of its antecedents is presented. The rest of the Chap. 2 concentrates on discussing each of the factors presented in the taxonomy in the light of existing literature. The aim is to consider the most relevant factors related to perceived location privacy. This includes identifying which user-related aspects–such as attitudes and concerns–are relevant, and discussing technologies that enhance end-users' location privacy. Also the different types of privacy behaviours are presented.

In Chap. 3 development of a questionnaire attempting at predicting privacy behaviour is presented. The development is based on the taxonomy postulated in Chap. 2, and considers the assumed direct antecedents to privacy behaviours, the *Location Privacy Beliefs and Attitudes* (LPBA). Additionally, development of a construct measuring location privacy knowledge is presented.

Chapter 4 dives into the topic of location privacy attitudes and beliefs in the context of location privacy through a qualitative study. The study concentrates mostly on risks and benefits that users perceive in the usage of LBS, attempting to answer the research questions 2.1, 2.2., and 2.3. The topic of trust is addressed, answering to RQ 1.3, and knowledge related to data privacy in LBS is discussed (RQ 1.4.).

In Chap. 5, users' location disclosing behaviour is studied. Two studies are conducted for the research: The first study presented in this chapter is a field study assessing location disclosure using a location-based mobile participation tool. In the study, the focus is on assessing what motivates users to disclose location information using such a tool. Then, a field study with a mobile messaging application prototype is presented. The study assesses the influence of a recipient (RQ 1.5.) as well as that of privacy concern (RQs 2.1., 2.3.), on privacy protection behaviour.

Chapter 6 assesses whether location privacy can be quantified. First, a crowd-sourcing study with various location disclosing scenarios is presented; the aim of the study is to evaluate users' location privacy valuation (RQ 2.6.). Within the study, the impact of a sharing scenario (RQ 1.5.), as well as that of trust towards the recipient of information (RQ 1.3.), on privacy attitudes is evaluated. Then, an online study is presented. The study tackles the question of how the monetary valuation of privacy relates to other factors presented in this work (RQ 2.6.). Also the impact of prior privacy violations (RQ 1.2.), as well as that of LBS knowledge (RQ 1.4.) on privacy attitudes, is assessed. The online study presented in this chapter also assesses the users' perceptions on privacy calculus: do the risks or benefits outweigh one another (RQ 2.2.), and does the privacy calculus influence behavioural intention (RQ 2.3.). Finally, the impact of social norm on behavioural intention is evaluated (RQ 2.4.).

In Chap. 7, privacy attitudes and beliefs are analysed with respect to how they relate to behaviour in terms of location disclosure and usage of privacy protection methods. The privacy attitudes in question are perceived risks and benefits (RQ 2.1.), risk–benefits assessment (or *privacy calculus*, RQ 2.3.), social norm (RQ 2.4.), perceived control (RQ 2.5.), as well as location privacy valuation (RQ 2.6.). Then, the connection between behavioural intention and behaviour is assessed (RQ 3.1.). Also awareness of behaviour is analysed in this context with the help of privacy nudges, aiming at answering to the RQ 3.2. Additionally, the influence of background factors on privacy attitudes and beliefs is evaluated; the focus here is on evaluating the influence of willingness to disclose, and that of trust, on privacy attitudes (RQs 1.1., and 1.3.). The research is conducted as a field study within which users have a location privacy protection application at their disposal, additionally, attitudes are assessed using a questionnaire.

Finally, the findings of this work are discussed in Chap. 8, together with an outlook to future work as well as recommendations for privacy research.

The following publications have been used in this work:

- Poikela, M., Schmidt, R., Wechsung, I., and Möller, S. (2014). Locate!—When do Users Disclose Location? In Workshop on Privacy Personas and Segmentation (PPS) at the Tenth Symposium On Usable Privacy and Security (SOUPS). Menlo Park, CA: USENIX Association.
- Poikela, M., Schmidt, R., Wechsung, I., and Möller, S. (2015). FlashPolling Privacy: the Discrepancy of Intention and Action in Location-Based Poll Participation. In PerPart 2015: 2nd International Workshop on Pervasive Participation, 2015 ACM International Joint Conference on Pervasive and Ubiquitous Computing (UbiComp 2015). Osaka: ACM.
- Poikela, M., Wechsung, I., Möller, S. (2015). Location-Based Applications-Benefits, Risks, and Concerns as Usage Predictors. In: 2nd Annual Privacy Personas and Segmentation (PPS) Workshop at the Eleventh Symposium on Usable Privacy and Security (SOUPS).
- Poikela, M., Schmidt, R., Wechsung, I., and Möller, S. (2016). "About your smartphone usage"—Privacy in location-based mobile participation. In International Symposium on Technology and Society, Proceedings (Vol. 2016–March). https://doi.org/10.1109/ISTAS.2015.7439421
- Poikela, M., and Kaiser, F. (2016). "It Is a Topic That Confuses Me"—Privacy Perceptions in Usage of Location-Based Applications. In European Workshop on Usable Security. Geneva: Internet Society.
- Poikela, M., and Toch, E. (2017). Understanding the Valuation of Location Privacy: a Crowdsourcing-Based Approach. Proceedings of the 50th Annual Hawaii International Conference on System Sciences.

Chapter 2
Theoretical Background to Location Privacy

2.1 Theories and Models to Privacy Behaviour

The relationships of behaviours and antecedents leading to such outcomes have been assessed in various behavioural theories. Such theories stem from different fields of behavioural psychology, and are used to explain and predict individuals' behaviour. Some have also been adopted—and adapted—for the field of privacy [102, 121], including location privacy [183]. The theories most influential on this work are presented within the following sections.

2.1.1 Theory of Planned Behaviour

The theory of planned behaviour (TPB) by Fishbein and Ajzen [58] suggests that various beliefs or attitudes affect behavioural intention. Subsequently, the intention is considered to be a direct antecedent to the behaviour in question. According to the theory, the possible positive and negative outcomes are first weighed. Second, the social norms—the social expectations of the peers with respect to the behaviour— are evaluated. Third, the perceived behavioural control is assessed. These factors influence the intention to engage in the behaviour, and through that intention, they affect the behaviour itself.

2.1.2 Theory of Reasoned Action

Theory of reasoned action (TRA) can be considered an update of the theory of planned behaviour. The theory is based on the TPB, and aiming at diminishing the

© Springer Nature Switzerland AG 2020
M. E. Poikela, *Perceived Privacy in Location-Based Mobile System*,
T-Labs Series in Telecommunication Services,
https://doi.org/10.1007/978-3-030-34171-8_2

discrepancy between attitude and behaviour, Ajzen introduces the actual control that user has over behaviour. The added component of actual control influences intention through perceived control, and also directly behaviour [9].

2.1.3   Reasoned Action Approach

Basing on the TPB and TRA, Fishbein and Ajzen further updated the model with various background factors, including individual, social, as well as information factors. Additionally, a feedback loop was added from behaviour to behavioural attitudes, which are the antecedents to behavioural intention. This framework is known as the reasoned action approach (RAA) [59].

2.1.4   APCO Macro Model

In the context of privacy behaviour, Smith et al. present a model based on previous literature assessing the relationships between privacy and other constructs [149]. The model combines the antecedents, privacy concern, and outcomes, and is therefore called the APCO Macro model. As antecedents, they have listed privacy experiences (here referring to negative experiences), privacy awareness, personality, demographics, and culture or social climate. These factors are suggested to directly influence privacy concerns, which include beliefs, attitudes, and perceptions. Also included in the model are regulations, privacy notices, trust, and privacy calculus, which is influenced by perceived risks and benefits. The risks, or privacy costs, have been suggested to be influenced by privacy concerns. The privacy concerns and regulation have a two-way relationship, as well as trust and privacy concerns. Trust is, according to the model, influenced by privacy notices. Finally, behavioural reactions are influenced by privacy concerns, by trust, and by privacy calculus.

2.1.5   Technology Acceptance Model (TAM)

Drawing from the TPB, Davis proposed a technology acceptance model (TAM), which states that acceptance of a technology is influenced by the perceived usefulness, and perceived ease of use of that technology [42]. These two factors influence the user's attitude towards using the technology, which in turn has an influence on intention to use, and through that, on actual usage of the technology.

2.1.6 Unified Theory of Acceptance and Use of Technology (UTAUT)

Acceptance of information technology has been studied by Venkatesh et al. in their proposed model named unified theory of acceptance and use of technology (UTAUT) [167]. The model is one of the most notable extensions of the TAM [42], and includes also various other behavioural models. The UTAUT assesses the intention to adopt new technologies, and comprises four key elements: performance expectancy, effort expectancy, social influence, as well as facilitating conditions.

Zhou examined the intention to use location-based services from the perspective of the UTAUT, and included privacy risk in the model [184]. They find that intention to use LBS is positively influenced by performance expectancy, and negatively influenced by privacy risks.

2.1.7 Expected Utility Theory

In situations involving risks, behaviour could to some extent be predicted using the expected utility theory, which was proposed by Bernoulli already in 1738. According to the theory, in situations where individuals are facing possibly risky outcomes, the highest expected value is not necessarily chosen, but in particular risk-averse individuals rather choose safer options to minimize losses [108].

2.1.8 Prospect Theory

According to the prospect theory by Kahneman and Tversky [87], individuals make decisions that involve risks based on the estimated gains and losses associated with the decision. Central to the prospect theory is that gains and losses are valued differently. According to the theory, individuals have inconsistent preferences among choices depending on how the choices have been presented. Included in the theory is the *endowment effect*, according to which individuals attribute more value to things that they own [86]. Such difference between the willingness to accept a price (WTA) for a reduction in privacy levels and willingness to pay for acquiring more privacy (WTP) has been also identified in practice, with WTA exceeding WTP in situations involving information disclosure [66], and online adverts [105].

2.2 Taxonomy

Building on the prior behavioural models, a location privacy taxonomy (LPT) is proposed (cf. Fig. 2.1). The taxonomy provides a foundation to assess location privacy, and to predict privacy behaviour in the context of location information.

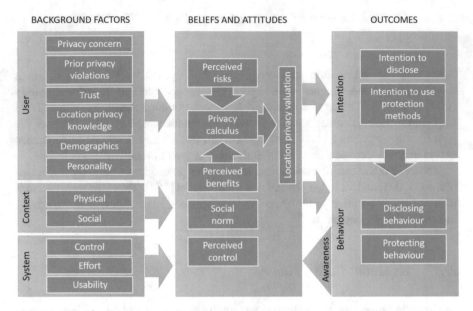

Fig. 2.1 The proposed taxonomy for location privacy (LPT), including background factors, beliefs and attitudes, and outcomes. The taxonomy, building on behavioural models, addresses behaviours as direct outcome of attitudes, which in turn are influenced by user-, context-, and system-related background factors. Intention and behaviour are addressed separately, and the path from intention to behaviour is also evaluated. Within beliefs and attitudes, perceived risks and benefits are combined into privacy calculus, which is assessed also in terms of monetary quantification; all beliefs and attitudes are addressed as influencing factors for behaviour. Finally, a feedback loop from behaviour to attitudes denotes the influence of behavioural awareness on beliefs and attitudes

The LPT taxonomy is a model aiming at predicting privacy behaviour in the context of location information, and includes three layers: the background factors, the beliefs and attitudes, and the outcomes. The layers of the model are presented in the following sections, with a description of each factor within the three layers provided.

2.2.1 Background Factors

This section discusses the background factors influencing location privacy intentions and behaviours mediated by beliefs and attitudes (cf. Fig. 2.2).

This first layer of the LPT, background factors, has three main building blocks: factors relating to *User*, *Context*, and *System* [172]. These building blocks are discussed here in more detail.

1. User-related factors, including

Fig. 2.2 The first layer of the
LPT presents background
factors to location privacy

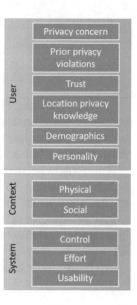

(a) Privacy concern, which in this work is considered to be a context-independent trait. It is expected to give a measure for the user's disposition to be concerned of information disclosure and influence the inclination for risk perception, but not necessarily provide an accurate direct estimate for privacy behaviour [80].

(b) Prior privacy violations such as uncomfortable information disclosures, or severe privacy breaches.

(c) Trust towards the recipient of disclosed information or towards the service provider.

(d) Location privacy knowledge, measuring the user's general understanding of data flows and companies' data privacy practices according to the current laws.

(e) Demographics, including age, gender, education, and occupation, and

(f) Personality, measured using the Big Five personality inventory [65].

2. Context, including

(a) Physical context including among other aspects the time and location, the thermal, visual, and auditory environment, and whether the user is indoors, outdoors, or in a vehicle. Context varies based on whether the user is, for example, at home, at work, commuting, or shopping, as the functions and roles that the user has in these different locations vary.

(b) Social context, including whether the user is accompanied by friends and family members, by strangers, or if they are alone. Social context can also be considered to extend to the individuals at the receiving end of location disclosures in social sharing situations. Also the social climate and culture can be considered part of the social context.

3. System factors, including

    (a) Control that the user has when using the system
    (b) Effort that it takes to use the system, and
    (c) Usability of the system.

These background factors are expected to influence the factors in the second layer—the beliefs and attitudes, and through that, have an indirect influence on the behavioural outcomes. In this book, the emphasis is on the first building block of the background factors, the user-related factors.

There might be various interdependencies between these factors. For example, experienced privacy violations could have an impact on privacy knowledge, and on the other hand, increased knowledge might give users more control over behaviour. Also, social climate—including regulations and public discourse—might influence trust. However, these interdependencies are not addressed in detail within this work.

All factors included in first layer of the LPT are presented here in detail.

2.2.1.1 Privacy Concern

Privacy concern can be considered a personal trait, which directly influences the user's beliefs and attitudes with respect to disclosing information. Earlier studies have found a connection between privacy concern and intention to disclose information [18, 102, 117]. Privacy concern has also been found to be a factor in whether or not a location-based technology is adopted [178]: users have initially privacy concerns regarding LBS, but after a while of using them, the concern diminishes [20]. Zhou et al. found that privacy concern does not affect intention to use LBS directly, but affects perceived risk, which in turn has a negative effect on usage intention [184]. Together perceived risks and privacy concern are expected to cover the contextual and context-independent aspects inhibiting privacy behaviour.

Users are often categorized based on their level of privacy concern, such as the Westin's privacy indices, which categorizes users based on their propensity to disclose personal information [95], or the categorization of users based on information privacy concern by Sheehan [147]. To measure privacy concern with respect to organizational practices, Smith et al. developed a construct measuring *concern for information privacy* (CFIP) [150]. To move from offline to online environment, Malhotra et al. updated the tool for *Internet Users' Information Privacy Concern* (IUIPC) [102]. To take the special issues related to mobile devices into account, Xu et al. created a measure for mobile users' privacy concerns [177]. Measuring concern on a global level while behaviour is very specific can also cause privacy behaviour to seemingly diverge from the privacy attitudes [80]. Thus, more specific or context-related measures for concern might be more accurate.

2.2.1.2 Prior Privacy Violations

Privacy violations are incidents where the user experiences discomfort because of the way that information about them has been interpreted, shared, or used beyond the originally intended purposes. Information shared online by one's friends [73], or even by oneself [170], can cause privacy boundary turbulence and discomfort [128]. Known risks and past experiences affect usage: past privacy violations impact individuals' privacy concern [150], and have the consequence for an individual to tighten their privacy protection mechanisms [156]. Privacy breaches can get costly also to the affected companies if customers' data has leaked to unintended entities [5, 131].

2.2.1.3 Trust

Mayer et al. defined trust as the willingness of a consumer to be vulnerable to the actions of a mobile service provider based on the expectation that the mobile service provider will provide services important to the consumer [103]. Mistrust in service providers creates agitation in users [34]. Building trust is a good strategy for the providers of LBS: trust in the receiving entity mitigates concerns of privacy risks [92, 149], and increases the users' willingness to engage in using LBS [179]. Transparency and good information privacy practices lead to enhanced trust [38], and to willingness to disclose information [80]. Trust has been found to influence privacy attitudes and privacy behaviour also in the context of location privacy: how willing one is to disclose location in various situations is influenced by who the requester of information is and how close to the recipient of the data the user feels [35, 166]—possibly as a consequence of higher trust. Trust can be positively influenced by information quality, system quality, and service quality [168]. Trust was found to have a negative relationship with privacy concern, and a positive relationship with LBS usage intention [168, 183].

2.2.1.4 Location Privacy Knowledge

Understanding information privacy practices when using LBSs can be considered crucial for being in control of the use, as well as subsequent use of one's information; however, the users are often unaware of the collection of location data [99], and uncomfortable when confronted with the information of to what extent data of them has been collected [148].

While the data collection might have been completely according to the data privacy laws and even stated in the privacy policies that the user has explicitly accepted, if the users' expectations of how their information is being handled are not met, the user often feels that their privacy has been violated [116]. In a study assessing internet users' knowledge of data privacy practices on the internet, the users were found to have vast knowledge gaps on how, if at all, data protection laws protect them [164]. Furthermore, users have been found to have very limited understanding of online privacy regulations [49], behavioural advertising [105] as well as locational privacy knowledge [10, 123]. The online users in the USA have been found to have an unfounded belief that laws and regulations protect their data from being passed on to third parties [17, 164], and also—mistakenly—believe that online stores cannot offer differing prices to customers based on the collected information. Users also drastically underestimate how much their data is used for different purposes [17]. A better understanding of information privacy practices seems to be correlated with higher privacy concern [72]; however, even when the users have a relatively good knowledge of data protection strategies, this knowledge does not seem to translate to actions towards data protection [49].

Merely providing a privacy policy cannot be considered a sufficient solution to inform the users. First, because users have been found not to read privacy policies [80], second, because it has been suggested that privacy policies are written in a language that is incomprehensible to the common user [79], and lastly, privacy policies are so lengthy that it would not be economically wise to read them [79]. These might, in addition to limitations in technical knowledge, be some of the main reasons why users do not understand the information flows of the services they use [164]. To overcome the issue, the GDPR, which came into effect in the EU in May 2018, requires companies to ask the users for specific consent for each data collection purpose in plainly worded text [160].

2.2.1.5   Demographics

Demographic differences are found in privacy perceptions. Differences were found in privacy attitudes of users with varying levels of education; users with higher education seem to have higher privacy concern about online privacy [147].

Some age-related differences have been found in privacy attitudes: older users are either strongly concerned about their privacy or not at all, while younger users seem to fall into the category of privacy pragmatists [147].

There are some gender differences in privacy perceptions; women are more concerned than men about information collection [38], and subsequent data use [146]. No gender-based differences were found in mobile privacy knowledge and skills [123].

2.2.1.6   Personality

Personality traits have been found to have an influence on privacy attitudes and behaviour: agreeableness, openness to experience, and conscientiousness influence intention to adopt LBS, mediated by privacy concerns [18, 83]. Privacy differences have been also found between introverted and extroverted users [101].

2.2.1.7 Physical Context

Physical context has an influence on user's privacy perceptions [81]. Physical context refers to aspects around the user, such as whether the user is indoors or outdoors, or in a possibly moving vehicle. Also environmental factors including the temperature, illumination, and noise can be considered part of physical context.

When talking about information privacy, Nissenbaum defines context as the consistency of data disclosure in such a way that users can assume their information not to be shared beyond the intended purpose [116]. This means that, for example, when a user makes purchase online, they can assume that any information from that transaction is not used beyond that purpose. Based on such a view, users can be presumed to behave in a way that is consistent with their privacy preference corresponding to each context.

2.2.1.8 Social Context

Social context includes whether the user is accompanied, and by whom; when the user is surrounded by crowds, there might be more concern for shoulder surfing or other privacy infringements. When in the presence of friends and family the user might also behave in a different way, mediated by social norm (cf. Sect. 2.2.2.4). Deriving from the definition of Nissenbaum [116], assuming that users alter their behaviour corresponding to different online situations, it can be postulated that the view also extends to the context changes because of different recipients of information disclosures. Individuals' privacy perceptions have been found to vary based on the social context [81].

Also other societal factors contribute to social context. These include political climate, whether there have been some big privacy incidents in the society, and news coverage; however, the impact of the news coverage might be only on short term [140]. Also privacy regulation might have an impact on users' privacy perceptions [149].

Social context is strongly connected to the culture in which the individual operates. Previous research indicates that individuals' privacy perceptions vary across cultures [25, 47]. In all cultures individuals try to control their privacy boundaries; however, the methods might vary from one culture to another [11]. The influence of culture is not assessed further in this work.

Social context can refer also to social expectations—these are assessed within the beliefs and attitudes as *social norm*.

2.2.1.9 System-Related Factors

Various system factors might potentially influence users' behaviour, assumably mediated by beliefs and attitudes. Such factors include the actual control that the user has when using the system, the effort of using the system, and the usability of the system. Technological context has an influence on individual's privacy perceptions [81].

According to the TRA, control that the user has over conducting a task using the system influences the perceived control, through which it also influences behaviour; control is thus considered a system factor.

System-related factors are studied in this work only to a limited extent. In the studies presented in Chaps. 5 and 7, participants are provided different levels of privacy protection mechanisms; however, the influence of having the control options is not systematically studied in comparison to not having the controls. The influence of effort and usability should be addressed in a comparative study where the users are provided with systems with varying levels of effort and usability.

2.2.2   Beliefs and Attitudes

In this section, the factors in the second layer of the LPT are discussed. This layer includes the user beliefs and attitudes: perceived risks as well as benefits, privacy calculus, social norm, perceived control, as well as location privacy valuation (cf. Fig. 2.3).

As stated in RAA, the behavioural intention is mediated by several attitudes towards the behaviour [59]. In RAA, first the positive and negative outcomes are weighed—this corresponds to evaluating the benefits and risks of using LBS. In the taxonomy, the perceived benefits and perceived risks are assessed separately, and the risk–benefit assessment, or privacy calculus, is then done to evaluate whether the risks or the benefits are considered more prominent. A previous work assessing

Fig. 2.3 The second layer of the taxonomy assesses the beliefs and attitudes related to location privacy

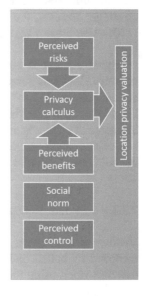

privacy behaviour in the context of location suggests that privacy behaviour is influenced by the perceived risks, but not by inherent, context-independent privacy concern [138]. This could be because of the privacy concern being measured on a rather general level, therefore not corresponding well with the context-specific privacy behaviour. Thus, privacy concern is in this work measured only as a background factor and possibly influencing the outcomes only mediated by beliefs and attitudes.

Next, the social norms, being the social expectations around the behaviour, are evaluated. Additionally, the subjective control that the user has for performing the behaviour aimed at reaching the intended outcome is included in the model.

Finally, the privacy calculus is expected to result in a quantification of privacy, within which a user attributes a certain value to their location privacy. This valuation, which is operationalized as monetary valuation, is used as a mediator for behavioural outcomes—for intention, disclosing behaviour, as well as protecting behaviour.

2.2.2.1 Perceived Risks

The risks that might entail from using LBS could be privacy violations that the user has unknowingly agreed to when accepting the terms of service, such as secondary use in terms of location-based advertising, or even more severe, such as theft or identity theft. Considering perceived risks, the users have been found to have particular worries when using mobile devices. These worries include physical damage, data loss, battery life, and lack of trust [34]. Other likely risks the users perceive in the usage of LBS include revealing one's home location, and being stalked [163]. Users are justified in being concerned, as sensitive information, including home and work locations, can be inferred from disclosed location data [61].

Also behavioural advertising that has overly precise targeting creates privacy concern in users, and as a consequence, may lead to decreased disclosure [62]. In such a case, the observer effect–the panopticon–of online media can be considered to be in effect, and inhibiting behaviour.

How risky a certain information disclosure is perceived to be is deemed an antecedent to privacy behaviour, while taking into account the context of the disclosure as well other background factors.

2.2.2.2 Perceived Benefits

The benefits perceived in sharing one's location using a LBS might be an improved service, enhancing one's social status, or making it easier or faster to communicate with others and search for information. According to findings from previous studies, reasons for location sharing range from being in contact and meeting friends, discounts and rewards [124], to promoting oneself by giving an interesting image of oneself [124, 158]. Also localized searches such as finding services nearby or finding information about a currently visited location are reasons to use LBSs [61]. Tsai et al. suggest that the biggest benefits of these services seem to be related to security (finding people in an emergency, or tracking the children in one's family), as well as finding information based on one's location [163].

2.2.2.3 Privacy Calculus

When using different technologies, the user needs to assess the benefits and risks, and based on this risk–benefit analysis decide whether or not to disclose personal information by using the service [57]. Sharing one's location using any LBS can be considered a transaction where the expectations of the received benefits are assessed against possible risks entailed in such a transaction, and an evaluation is made whether or not it is reasonable to use the service. Users may not be happy about the perceived risks, but accept them as a compromise in order to receive the benefits [7]. This phenomenon where individuals accept a privacy risk in order to receive benefits is known as privacy calculus [48]. However, the privacy calculus does not always lead to the user to conduct in an optimal way, as users have a tendency to excessively value immediate benefits, outweighing them over possible future risks [2]. According to Dinev et al., perceived benefits often exceed the perceived risks while making disclosing decisions [48]. On the other hand, Kransnova et al. found that on online social networks, perceived risks often overpower perceived benefits, leading users to restrict their disclosing behaviour [92].

In practice users mostly do not show perfect judgement in their privacy decisions, a phenomenon presented in Sect. 2.2.5. Despite high privacy concerns, users still share their personal information in many circumstances for gained benefits. Furthermore, to be able to adequately assess the relationship of benefits and risks, the user needs to have a clear understanding of who collects the information, what the information is collected for, how long it is stored, and what are the possible secondary uses of it; however, users often have knowledge gaps and misunderstandings when it comes to privacy (cf. Sect. 2.2.1.4).

2.2.2.4 Social Norm

Social norms refer to how an individual believes that others expect them to behave. In the context of privacy, this could be either regarding disclosing behaviour using location-based applications, or protective behaviour, such as obfuscation. Social norms have been found to influence protection behaviour on social networking sites: beliefs that one's peers protect their private information on SNSs was found to be connected with more use of restrictive settings [165].

2.2.2.5 Perceived Control

Perceived control in this work refers to the control that a user perceives they have over performing an intended behaviour.

Users often have little to no control over what happens to their information after it has been shared. While the data collection might have been completely according to the data privacy laws and even stated in the privacy policies that the user has explicitly accepted, if the users' expectations of how their information is being handled are not met, the user feels that they have lost some control over their information disclosure, and their privacy has been violated [116]. Individuals' information privacy concern can be alleviated by raising their perceived control over personal information [180].

Four different scenarios can be speculated around the relationship of the feeling of being in control of one's location privacy and the usage of methods to protect one's location privacy:

(a) The user feels being in control because privacy protection mechanisms are available (and the user knows how to use them).
(b) The user feels that they are in control and do not need further protection; therefore, the user does not use privacy protection methods.
(c) The user feels that they are not sufficiently in control because of lacking privacy control mechanisms or the knowledge of how to use them.
(d) The user feels that they are not sufficiently in control, and therefore use privacy protection methods. In an optimal case, the situation evolves through the privacy protection mechanisms into the scenario *a*.

2.2.2.6 Valuation of Location Privacy

As noted earlier in Sect. 2.2.2.3 in the context of privacy calculus, using location-based applications can from a user's perspective be considered a trade-off, where benefits are gained at the expense of a risk to one's privacy. Thus, location privacy can be considered a commodity that the user exchanges for services. This leads to the conclusion that the user must define a value for the location information prior to sharing.

Location disclosures can typically be measured on a binary level—the user either shares location or does not. The privacy calculations leading to the decision, however, are more complicated than such binary simplifications [4], and the user has to decide whether or not the benefits received from using a LBS are worth the possible risks. Psychological behavioural theories suggest that behaviour is preceded by behavioural intentions [9]; however, for assessing privacy behaviour, it is not sufficient to consider intention, as there is a discrepancy between disclosure intention and actual behaviour [117]. Using monetary valuations as a proxy for privacy provides a tool to evaluate the user's privacy perceptions with greater granularity, where the valuation can be considered a quantification of privacy calculus.

There are contradicting findings in the privacy literature as to how users value their privacy, for example, whether the consumers are willing to pay for protective mechanisms [117] or not [6]. The value that users give to their privacy has been studied in various offline and hypothetical online situations, with large variations found based on the context [7].

The amount of money that users are willing to pay for protecting otherwise public personal information is significantly smaller than the amount that the same users would require for disclosing otherwise private information [7, 104], a phenomenon known as the endowment effect [85]. This endowment effect, as well as the order in which the user is presented with the alternatives, influence privacy valuation [145]. Privacy valuation seems to vary between individuals and based on a context; the value users attribute to location privacy seems to be the highest for the users with most variance in their movement patterns [40], and the lowest at locations with large and diverse sets of visitors [161].

2.2.3 Behavioural Outcomes

In RAA, mainly the intention to behave is evaluated as an outcome factor, and intention is expected to directly lead to the corresponding behaviours [59]. In the LPT taxonomy presented in this work, the outcomes are divided into two categories: behavioural intentions and privacy behaviours, both of which are further divided into the aspects of disclosure and usage of protection methods.

This section discusses the third layer of the LPT, including behavioural intentions as well as behaviours in the context of location privacy (cf. Fig. 2.4).

Behavioural intentions in the context of location privacy refer to the intention to engage in the usage of LBS in the future, to disclose location information by other means including social media, or the intention to use privacy protection mechanisms.

Privacy behaviour is divided into location disclosing behaviour and protective behaviour, where a PET or other location privacy protection mechanism is in use. These concepts can appear also in parallel.

This section also briefly discusses the location privacy-enhancing technologies that users have at their disposal, as well as how there sometimes seems to be a discrepancy between the privacy attitudes and behaviour, a phenomenon known as the *privacy paradox*.

2.2.3.1 Behavioural Intention

In the LPT taxonomy presented in this work, two types of behavioural intentions are identified: the intention to disclose location, and the intention to protect oneself from

Fig. 2.4 The third layer of
the LPT includes behavioural
intentions and behaviours in
the context of location
privacy

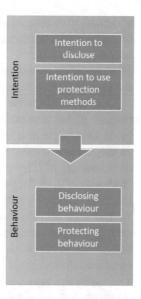

privacy risks by using protection mechanisms. In the privacy literature, behavioural
intention is often measured as the only outcome variable, assuming that it would
correspond to actual behaviour as postulated in the TPB and the TRA [149]. Also
studies stating that privacy might play only little role in intention to use online
services have been reported [24].

Intention to disclose refers to the intention of a user to use location-based
applications and thereby disclose their location information to different entities, or
intention in other ways to disclose their location information, such as revealing it
explicitly in social media. Such behaviours are referred to as disclosing behaviours.

Intention to use protection methods refers to the extent to which a user plans to
use methods such as avoiding usage of LBS, uninstalling applications, switching
location services off, or using some other methods, or privacy-enhancing technologies (PETs) for location privacy. Intention to engage in protecting behaviour, and
the relationship of such intention and the protecting behaviour, are not assessed in
this work.

2.2.3.2 Location Disclosing Behaviour

Self-disclosure is the telling of the previously unknown so that it becomes shared
knowledge, the "process of making the self known to others" [78]. Self-disclosure
through online platforms has several purposes, such as letting others know one's
opinions by writing posts and writing blogs, thereby differentiating oneself from
others; fulfilling the "need to be seen" by reminding others of one's existence; or
maintaining personal life and a public face [82]. Users maintain an online presence,
showing a picture that they are willing to communicate to others of themselves, and

which might be filtered to portray a favourable image of oneself [82]. Often the users of social media platforms protect their privacy by managing their lists of contacts and the access rights that people on these lists have to personal content [181]. However, they might not consider the amount and detail of personal information the providers of these services collect of them.

Other reasons to disclose personal information include security, for example, through giving personal information for authentication and safety through surveillance. One might also be willing to disclose more information for affiliation as well as allegiance purposes: for enhancing the bonds of trust between group members, or for increasing trust by making oneself vulnerable through self-disclosure (e.g. "I am an alcoholic"), or for personal growth within therapeutic sessions [127]. Scientific purposes might also motivate users to disclose personal information.

The term self-disclosure is often associated with interpersonal intimacy; however, it does not necessarily mean disclosing private information, as the information that is shared between individuals is not necessarily personal. However, more sensitive information is shared with emotionally closer acquaintances [119].

Despite the possible privacy concern, often social, economic, or other benefits outweigh the risks, resulting in information disclosure [38]. Users of LBS have various reasons to disclose their location information, such as being in contact and meeting with friends, commercial benefits and rewards [124], promoting oneself, or presenting oneself in a better light to others [124, 158]. The requester of location information is a remarkable influencing factor determining location disclosure—and seems to play a bigger role in the decision to disclose location information than current location of the user [97]. Users are practical in their location disclosures; location information is shared with the precision that is perceived as the most meaningful for the recipient of the information, rather than deciding on the degree of precision based on other parameters, such as privacy concern [35].

2.2.3.3   Awareness of One's Location Disclosure

To be able to adequately assess privacy risks, users need to be made aware of the possible risks that might be connected with their behaviour. The traditional notice and control approach, where companies merely provide a privacy policy leaving the burden on the user to understand and agree the given terms, is not sufficient for this purpose, as users do not read privacy policies [79]. To improve users' understanding of their privacy choices, Kelley et al. [89] suggested "nutrition labels" that would standardize privacy policies and make them easily accessible and comprehensible.

As noted by Smith et al., privacy notices and privacy seals influence users' trust, and through that, both the privacy perceptions as well as behaviours [149, 179]. Because privacy seals are assumed to have only an indirect influence on privacy perceptions, through trust [179], they are not taken into further consideration in this work.

Phelan et al. discuss that users might intuitively commit to privacy sensitive activities, but in particular when prompted, evaluate the privacy risks more rationally

[130]. However, when users are given feedback about their disclosures, without giving them any control for changing privacy settings, discomfort ensues [125].

Users are often unaware of much of the data disclosures on their devices, and wish they had more information available than currently is the case [16]. Transparent data privacy practices have been found useful in decreasing users concerns of surveillance [120]; however, contradictory findings have also been reported [10]. Informing users of LBS about their data disclosure prompts a re-evaluation, and leads to restricting permission given to applications [10]. This soft paternalistic method of informing the user of their behaviour is frequently called privacy nudges [17]. These nudges are interventions that inform the users, for example, about what possible consequences their behaviour might have, in order to guide them towards more privacy-conscious choices [3]. Reminding users about the audience of their disclosures can prevent unintended disclosures [169]. These nudges can be shown at set-up, just-in-time, depending on the context, periodically, persistently, or on demand [16, 144]

2.2.3.4 Location Protection Behaviour

Users have few tools at their disposal to protect their privacy while using LBS. These are shortly discussed here. First, users can inform and educate themselves, for example, by reading privacy policies or application permissions before installing smartphone applications. The application permissions is a list of features on the device to which a certain application has an access, including the users' contacts, calendar, and location [56]. Recent studies suggest that users might read these permissions prior to installing applications, but do not understand them and are unaware of the associated risks [90]. However, users can be directed towards making better informed decisions through a better presentation of the permissions [90, 94].

Second, to protect one's privacy, a commonly used risk mitigation technique is to avoid using smartphones or LBS altogether, as well as uninstalling applications [133, 136]. To avoid LBS from collecting location data, users might switch the location services off altogether, thus also not being able to use features that use the location of the device. A common misunderstanding that the users have is that switching the GPS off prevents the device's location from being used [133].

To manage location privacy settings with more precision, a user might also use location obfuscation. Some prototypes have been suggested, giving users a possibility to adjust the location precision shared with others according to their privacy preferences [134]. Such a functionality is not, however, readily available in most systems to date. In Apple's iPhones, a user can give applications a permission to use the device's location either only when the application is actively used, when it is in the background, or never. On Google's Android smartphones, a user can set the location sharing on or off for each application. However, the user has no control over how the location information is used once shared, and whether it is stored for some possible future use. The user has very limited possibilities of making sure what purposes their information is being used for, and often the user is not aware of how frequently their location is shared [10].

2.2.4 Privacy-Enhancing Technologies for the Context of Location

Privacy by notice—or so called "notice and choice"—is a method of taking care of end-user privacy by effectively leaving the burden to the user to read the privacy policies and understand what has been agreed on. This method has been widely criticized, and according to Spiekermann and Cranor, does not provide sufficient privacy protection [152]; instead, they suggest to use approaches of implementing so called "privacy by architecture".

End-users' privacy is sometimes protected by anonymizing the data with pseudonyms. The pseudonyms replace the personally identifiable identification such as names or personal identification numbers with an untraceable ID. These can be either long-term, or as suggested by Beresford and Stajano [28], change frequently in order to make it less likely that an attacker could collect enough information to infer anything about the data.

Encryption of communication has been used as a method to protect end-users' location privacy from malicious third-party attacks. However, encryption does not protect from server-side attacks [84]; *location obfuscation* has been suggested as a remedy [84]. Duckham and Kulik define the method of location obfuscation as deliberately degrading the quality of location information in order to protect one's privacy [50]. This process can be done, for example, by adding random noise to the location information [13], also known as *perturbation*, or by using spatial generalization [68]. It is however possible to reveal the identity of a device in the cases of continuous location tracking, i.e. where a user moves and their device repeatedly reports the location to a server [132]. Brush et al. investigate users' preferences for location obfuscation; different types of obfuscation methods were preferred, which were found to be in line with the individuals' privacy concerns [32].

Location k-anonymity is a concept stating that a user is k-anonymous if they cannot be distinguished from a location dataset of at least $k-1$ individuals [68, 157]. There are two common methods for achieving k-anonymity: suppression of parts of the data, and generalization, where an exact location is replaced by a larger area or a general landmark [157]. The concept of *differential privacy* states that a person's privacy cannot be compromised if their information is not in a database; thus, the person should be provided with approximately the same amount of privacy as they would have if their information was removed from the database [51]. The technique requires relatively large datasets, and otherwise requires more noise to be added, which in the case of location data might render the information unusable. Solutions to provide differential privacy for LBS have been proposed e.g. by Bordenabe [29].

2.2.5 Privacy Paradox

Users of mobile phones and other technologies often state high concern over their privacy; however, it seems that there is a discrepancy between these stated concerns and their actual disclosing behaviour. Users often seem to be willing to give out personal information opposing the previously stated intentions [2, 135, 153], even with no compensation [141].

In practice users are not always showing perfect judgement in their privacy decisions, but tend to excessively value immediate benefits, outweighing them over possible future risks, a phenomenon called *hyperbolic discounting* [2]. This phenomenon typically manifests as behaviour that seems irrational; a user might state high privacy concern but still disclose personal information for a benefit that appears very small. This discrepancy in privacy attitudes and behaviour is known as privacy paradox [117]. On the other hand, if users are unaware of how information is collected, where it might land, and the ways it might finally be used, also protecting oneself from privacy breaches and assessing the possible consequences becomes unlikely. Furthermore, according to Acquisti and Grossklags [6], users' decision process is affected by bounded rationality, whereby the rationality of the decision-making is limited by various factors, including cognitive limitations and lack of time. Therefore the behavioural intentions cannot be assumed to perfectly correspond with the behaviour.

2.3 Chapter Summary

This chapter provides an overview to literature related to location privacy. First, some of the most important behavioural models are presented to provide a foundation for a taxonomy of antecedents and outcomes of location privacy. The LPT taxonomy is presented with three main layers, with each factor discussed in the light of existing literature.

The first layer assesses the background factors, including user-related, context-related, as well as system-related factors. The system considers control, effort, and usability. In the context, physical and social contexts are considered. The user-related factors are location privacy concern, prior privacy violations, trust, location privacy knowledge, as well as demographics and personality. Privacy concern and trust are, unlike in some other works, considered as traits, and therefore included in the background factors rather than the second layer, which considers beliefs and attitudes.

The beliefs and attitudes include the context-related perceived risks, benefits, social expectations, and controls. The risk–benefit calculation, or *privacy calculus*, is presented as a factor being influenced by the perceived risks as well as the perceived benefits. Privacy calculus is postulated to influence the value that a user attributes to their location privacy.

The third layer of the LPT includes outcomes that are influenced by the previous layers—location privacy intentions and behaviours. The intentions as well as the behaviours are postulated to be influenced by the various dimensions of the second layer. Additionally, as suggested by RAA, if a user is made aware of their behaviour, this awareness might influence the beliefs and attitudes corresponding to that behaviour.

The taxonomy is a foundation for the rest of the book, and the relationships postulated therein are assessed in the subsequent chapters.

Chapter 3
How to Predict Location Privacy Behaviour?

3.1 Development of the Questionnaire for Location Privacy Beliefs and Attitudes (LPBA)

To predict privacy behaviour in the context of location as postulated in the taxonomy in Sect. 2.2, a measurement instrument was constructed. The final instrument is a questionnaire based on the second layer of the taxonomy, comprised of beliefs and attitudes. The first layer of the taxonomy is addressed only partially in the questionnaire development; in particular the user-related variables are measured and their relationship to the second layer is evaluated through regression analysis. The development of the constructs included in the first layer is presented in the Sect. 3.2.

The instrument was constructed during various studies and improved over the course of this research. The final instrument consists of three scales, and the construction of each is presented here. This chapter is largely following the methods reported by Wechsung [172] and Naderi [113].

3.1.1 Item Selection

Following the taxonomy presented in Sect. 2.2, a questionnaire to predict behaviour in the context of location privacy was created. Five dimensions for direct antecedents to privacy behaviour were identified, namely *Perceived Risks*, *Perceived Benefits*, *Privacy Calculus*, *Social Norm*, and *Perceived Control*.

Most answers are given on an end-labelled seven-point Likert scale from "Fully disagree" (0) to "Fully agree" (6), a method successfully adopted in previous works conceptualizing constructs [8]. The exceptions are mentioned separately. For an example of how the response choices were presented to the participants, see Fig. 3.1. This kind of response scale was chosen because reliability of responses is highest

© Springer Nature Switzerland AG 2020

M. E. Poikela, *Perceived Privacy in Location-Based Mobile System*,
T-Labs Series in Telecommunication Services,
https://doi.org/10.1007/978-3-030-34171-8_3

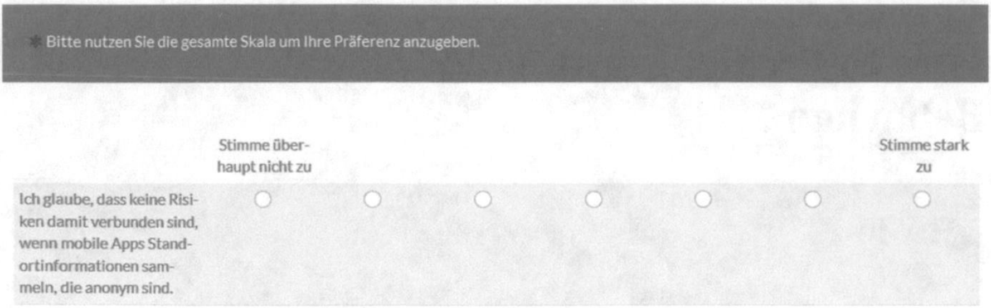

Fig. 3.1 An example of the used response scale

for a seven-point answer scale, and validity is highest for answer scale with five to nine response categories [55, 172].

First, for each dimension, an exploratory factor analysis (EFA) is conducted to evaluate whether all items theoretically assigned to a dimension form a single factor [33, 71, 172]. EFA is typically deployed when the underlying structure of correlations is not clear [33]. Unrotated Maximum Likelihood method was deployed.

At least 50% of the variance should be explained by each identified factor [55]. The fit between the hypothesized model and the data can be estimated using the χ^2 goodness-of-fit statistics, for which a non-significant χ^2 value can be considered an indication of a good model fit [33, 172]. These results are reported in Table 3.2. To calculate the χ^2 goodness-of-fit statistics, a positive degree of freedom is required, for which at least four items are needed. Thus, for constructs with three items or less, the χ^2 goodness-of-fit statistics could not be calculated.

Face validity of each factor was measured using a panel of expert judges. Face validity is defined as the extent to which the construct measures what it is intended to measure [118]. A group of 3–5 expert judges assessed how well each item represents the construct in question; not all items were assessed by the same judges, and not all judges assessed all the items. The judges rated each item either as "not representative", "somewhat representative", or "clearly representative" of the construct. To be deemed acceptable, an item is required to be "clearly representative" by at least 50% of the judges, and "somewhat representative" by the rest of them. In the following, in all cases where an item did not pass the requirement for face validity, it had already been removed because either it did not load on the main factor in the EFA or it deteriorated the internal consistency. Therefore, such cases are not reported in the following sections.

As an indication for internal consistency, Cronbach's α [37] was measured for each construct. Internal consistency refers to how closely related the items of a factor are to each other. A Cronbach's α value of >0.7 is recommended as a threshold [71], and where applicable, items were deleted to improve the measure.

In the following subsections, the generation and selection of initial set of items for each dimension is presented.

Risk Perception (RISK)

For measuring privacy concern in the context of location privacy, a set of items was created based on IUIPC [102], modified for the mobile context. The modifications included changing the terms related to online context with mobile context.

The first version of the construct had 42 items. All introduced items are listed in the Appendix A.9. The full set of items was presented to nine expert judges, who identified six factors within the dataset. However, one of the categories, representing *Notice and Disclosure* included only one item, and was therefore not considered; the item belonging to the category was removed. The remaining categories were named

- *Purpose and Secondary Use (PUSU)*
- *Unauthorized Access (UNAUT)*
- *Information Accuracy (ACCU)*
- *Data Collection (COLL)*, and
- *Perceived Risks (RISK)*, assessing particularly perceived risks in the context of LBS.

These five factors were subsequently confirmed using a card-sorting technique with six expert judges.

A series of three online studies was conducted to evaluate the identified factor structure (total $N = 98$). The studies are presented in the appendices (cf. Appendices A.3, A.4, and A.5). The EFA was conducted separately for each of the constructs. Items that were not loading on the main factor were deleted. Then, based on a reliability analysis, the items that deteriorated the Cronbach's α value were removed. The results for the EFA for the constructs related to risk perception are listed in Table 3.1.

The construct named *Perceived risk* was chosen to be used to measure perceived risks in the usage of LBS, as it was considered to most accurately reflect the intended dimension. For the subsequent studies, the other scales were disregarded.

Table 3.1 Results of the EFA and internal consistency of the constructs related to risk perception

Scale	Variance explained	χ^2	p	Initial N of items	Final N of items	Cronbach's α
PUSU	58.17%	32.21	0.906	13	11	0.936
UNAUT	55.14%	5.23	0.073	6	4	0.819
ACCU	73.88%	n/a	n/a	4	3	0.890
COLL	66.94%	n/a	n/a	3	3	0.854
RISK	62.80%	15.861	0.070	15	7	0.908

PUSU purpose and secondary use, *UNAUT* unauthorized access, *ACCU* information accuracy, *COLL* data collection, *RISK* perceived risks

Benefit Perception (BENE)

Based on existing literature, a set of seven items was created to assess perception of benefits in the usage of LBS. These items were deployed in a series of four studies to identify the factor structure (cf. Appendices A.3, A.5, A.6, and A.7). Users' benefit perception in the usage of LBS was further investigated within an exploratory interview study assessing the perceived risks and benefits (cf. Sect. 4.1). Based on the findings, five new items were generated, one was rephrased, and three were left out of further analysis.

The nine-item construct was deployed in an online study (cf. Sect. 6.2) and in a field study (cf. Sect. 7.1). A random sample (33.3%, $N = 59$) of the whole dataset was used for calculating the scale statistics; the remaining set was left to be used for validating the final model.

Various possible benefits were not included in the benefit perception scale, including receiving discounts, altruism, and stimulation. As a consequence, the construct has a limited content validity. Content validity refers to the extent to which the items in the measure represent an adequate sample of the theoretical content domain of the construct [118].

Privacy Calculus (PCAL)

For assessing privacy calculus, or the assessment of risks and benefits in the usage of LBS, one item was created. The question for measuring privacy calculus was worded as follows:

Q. *How do you assess the benefits and risks of using location-based services?*

The answers were collected on a single-choice scale with the following answer options:

- The benefits are much greater than the risks.
- The benefits are somewhat greater than the risks
- The benefits and risks are equal.
- The risks are somewhat greater than the benefits.
- The risks are much greater than the benefits.

EFA or internal consistency assessment could not be conducted for the one-item construct.

Perceived Norm (NORM)

Altogether 12 items were created to measure the peers' influence on the usage of LBS—the perception of how ones' peers are using LBS, and are expecting one to use them as well. These items were deployed in three studies to identify the factor structure (cf. Appendices A.3, A.5, and A.6). Based on the results from the EFA, three items were selected to represent the construct NORM (cf. Table 3.2).

Table 3.2 Results of the EFA and internal consistency of the constructs

Scale	Variance explained	χ^2	p	Initial N of items	Final N of items	Cronbach's α
Risk perception (RISK)	62.80%	15.861	0.070	15	7	0.908
Benefit perception (BENE)	60.3%	8.43	0.134	14	5	0.880
Privacy calculus (PCAL)	n/a	n/a	n/a	1	1	n/a
Perceived norm (NORM)	65.04%	n/a	n/a	12	3	0.803
Perceived control (CTRL)	42.13%	6.16	0.046	5	4	0.729

Perceived Control (CTRL)

Five items were created for measuring the perception of the control the user has in disclosing their location information. Similarly to Benefit Perception, the Perceived Control items were deployed in an online study (cf. Sect. 6.2) as well as in a field study (cf. Sect. 7.1); the same random sample (33.3%, $N = 59$) of the whole dataset was used for calculating the scale statistics as for BENE. One item was found to decrease internal consistency of the construct, and it also did not pass the face validity assessment. The EFA results for the remaining four items were not satisfactory: χ^2 goodness-of-fit statistic turned out significant, indicating an unsatisfactory model fit.

3.1.2 Factorial Structure of LPBA

The results for the EFA for the constructs in question (RISK, BENE, PCAL, NORM, and CTRL) are presented in Table 3.2.

The LPBA questionnaire under development was in parallel used in an online study (cf. Sect. 6.2), and in a field study (cf. Sect. 7.1). From the total dataset, a random sample (80%, $N = 137$) was used for training the data, thus leaving the remaining data for validating the final model.

Confirmatory factor analysis (CFA) was conducted to verify the factor structure using the sample data. This was done using IBM® SPSS® AMOS (version 25), which is a software package for structural equation modelling (SEM). SEM is a procedure to model multivariate relationships, with two main aspects: (a) the causal relationships are represented by a series of structural equations using regression and (b) the structural relationships are illustrated to provide a clear conceptualization of the theory [33].

The following model fit indices were used, with the thresholds recommended by Hu and Bentler [75], Homburg and Giering [71], and Wechsung [172] deployed:

The Ratio of χ^2 and the Degrees of Freedom (df); $\frac{\chi^2}{df} \leq 3$

Considering a null hypothesis, according to which the factor loadings, factor variances and covariances, as well as error variances for a model are valid, the χ^2 statistic gives an indication of whether or not the null hypothesis holds. A non-significant value is desired for a good model fit [33].

A large sample size negatively influences the statistical significance of the χ^2; to account for the issue, it has been suggested to adjust the χ^2 statistics by the degrees of freedom. This fit statistic is deployed in this work notwithstanding the fact that the sample size is likely not too high.

Comparative Fit Index (CFI) ≥ 0.95

The comparative fit index is an *incremental fit index*, comparing the hypothesized model with a *null model*, where all variables are uncorrelated [27]. Values range between [0,1], with a higher value suggesting a better fit for the hypothesized model.

Root Mean Square Error of Approximation (RMSEA) ≤ 0.08

As opposed to the CFI, the RMSEA is an *absolute fit index*, which is obtained without a reference model [33]. In RMSEA the error of approximation in the population is assessed, and smaller values indicate a better model fit.

The following thresholds are recommended by Hair et al. [70].

Composite Reliability (CR) > 0.7

Composite reliability measures the total reliability of a factor, indicating how well the factor is represented by the items loading on it [33]. The range for CR is [0,1], where a greater value indicates higher reliability.

Average Variance Extracted (AVE) ≥ 0.5

Average variance extracted is a measure of convergent validity, and is used for the global model for validating a questionnaire instrument. The measure gives the relation between the variance explained by the factor, and by a measurement error; a sufficient AVE value suggests that the items correlate well within the factor.

Table 3.3 Fit indices for the intermediate item set for beliefs and attitudes

Scale	N of items	$\frac{\chi^2}{df}$	p	CFI	RMSEA
Risk perception (RISK)	6	0.94	0.49	1.00	0.00
Benefit perception (BENE)	5	1.15	0.33	0.99	0.03
Privacy calculus (PCAL)	1	n/a	n/a	n/a	n/a
Perceived norm (NORM)	3	n/a	n/a	n/a	n/a
Perceived control (CTRL)	4	4.25	0.04	0.97	0.16

Discriminant Validity

Discriminant validity is a second measure that is used for assessing only the global model, and it is the degree to which a construct differs from dissimilar constructs. The requirement for discriminant validity is met during EFA when the correlation factor between constructs is at the most 0.7. During CFA discriminant validity is met when Maximum Shared Variance (MSV) < AVE, and the square root of AVE is greater than inter-construct correlations.

All constructs were assessed using the aforementioned criteria ($\frac{\chi^2}{df}$, CFI, RMSEA, and CR; cf. Table 3.3). Some of the constructs had three or fewer indicator variables; such constructs are *just identified* or *under-identified*. A construct that has three indicator variables is just identified, and when considering CFI or RMSEA for such a model, the fit is always perfect. Models that are under-identified do not have enough indicator variables (i.e. less than three) in comparison to the parameters to be estimated [172].

For the construct privacy calculus, the indices could not be calculated because it is under-identified, as it has only one indicator variable. Perceived norm is just identified as it has three indicator variables; the fit indices are therefore not meaningful. Finally, the results form perceived control suggest a bad model fit. To overcome the issue, modification indices computed by AMOS were investigated. Modification indices are suggestions for additional parameters, such as regression weights or covariances, that could be specified in order to improve the model fit [27]. A covariance between two error terms was added, which improved the model fit to some extent, however, the $\frac{\chi^2}{df}$, as well as RMSEA were still not satisfactory, thus the construct is left out of further analysis. After disregarding perceived control, fifteen items remain.

3.1.3 Validation of the LPBA Instrument

In the following development stage, the structure of the global model was evaluated using EFA [172]. A maximum likelihood factor analysis with a Promax rotation was conducted on the fifteen items. Based on the theorized factorial structure, four

Table 3.4 Pattern matrix for
the three factor solution for a
beliefs and attitudes based on
a Maximum Likelihood factor
analysis using Promax
rotation

	Factor 1	Factor 2	Factor 3
	RISK	BENE	NORM
RISK3	0.59		
RISK5	0.82		
RISK6	0.68		
RISK7	0.79		
RISK8	0.78		
RISK9	0.73		
BENE1		0.70	
BENE2		0.77	
BENE3		0.85	
BENE4		0.89	
BENE5		0.55	
NORM1			0.71
NORM2			1.03
NORM4			0.42

All factor loadings below 0.3 are suppressed
($KMO = 0.86$; $Bartlett : \chi^2(90, N = 133) = 963.48$; $p < 0.001$)

factors were extracted. However, the resulting factor structure was unsatisfactory.
Somewhat unsurprisingly, the one item for PCAL (PCAL1) loaded negatively on
the factor consisting mainly of BENE, and additionally, NORM4 did not load on
the main factor of NORM. Subsequently, PCAL1 was removed from the model, and
the factor analysis was conducted again with three factors extracted. The resulting
factorial structure is presented in Table 3.4. NORM2 can be seen to be a *Heywood
case* [91], which is here seen as a factor loading of greater than one. This might be
due to the sample size.

No cross-loadings were identified in the model. Also, there were no cross-
factor correlations above the recommended threshold of 0.7; the correlation between
BENE and NORM, 0.62, was the highest. Thus, the requirement for discriminant
validity was met. The three factor model explained 59.27% of the variance.

Next, the CFA using SPSS AMOS was conducted for the three-item model
deploying the criteria presented earlier. The model (cf. Fig. 3.2) provides a good
fit; $\frac{\chi^2}{df} = 1.23, CFI = 0.98, RMSEA = 0.04$. For each factor, AVE was above 0.5
indicating a good convergent validity. Additionally, for each factor $MSV < AVE$,
which represents an acceptable discriminant validity. The results are reported in
Table 3.5.

The remaining 20% ($N = 34$) of the dataset comprised of an online study and
a field study was used to validate the questionnaire. The fit indices presented in
Sect. 3.1.2 were used. The results do not suggest a good model fit: $\chi^2 = 95.51, p = 0.047$; $\frac{\chi^2}{df} = 1.29, CFI = 0.89, RMSEA = 0.09$. This result could not validate
the model, and a new data collection would be required in order to further evaluate

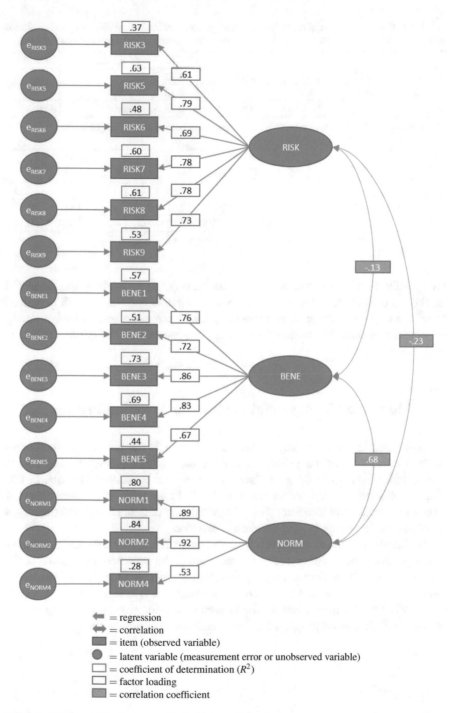

Fig. 3.2 CFA model for factors predicting privacy behaviour

Table 3.5 Composite reliability (CR), average variance extracted (AVE), maximum shared variance (MSV), square root of AVE, as well as inter-factor correlations for the training dataset

Scale	CR	AVE	MSV	RISK	BENE	NORM
RISK	0.87	0.54	0.05	**00.73**		
BENE	0.88	0.59	0.46	−0.13	**0.77**	
NORM	0.84	0.64	0.46	−0.23	0.68	**0.80**

The values depicting square root of AVE are presented in bold along the diagonal

Table 3.6 Descriptive statistics of the constructs: mean, standard deviation, correlations, and Cronbach's α

Scale	M	SD	Correlations		α
			RISK	BENE	
RISK	3.64	1.26	–		0.85
BENE	2.61	1.19	−0.14	–	0.87
NORM	2.97	1.06	−0.17*	0.57**	0.79

$^*p < 0.05; ^{**}p < 0.01$

the validity. Finally, the descriptive statistics using the full dataset are presented in Table 3.6. Risk perception has only a low negative correlation with perceived benefits and perceived norm. Perceived norm and perceived benefit have a strong positive correlation. All constructs have high internal reliabilities (Cronbach's α > 0.7).

3.2   Measuring Background Factors for Location Privacy

As postulated in the taxonomy presented in Sect. 2.2, the behavioural outcomes in the context of location privacy can be predicted from privacy-related beliefs and attitudes, which in turn are influenced by various background factors. In this work, mostly user-related factors are considered. In the following, development of a construct measuring location privacy knowledge is presented, and additionally, a construct for measuring privacy concern is assessed.

Also trust is addressed within this work, however, it is highly context-sensitive, as trust typically refers to either trust in the recipient of the disclosed information or the service provider. Therefore attempts at developing a construct including trust seem rather futile. The influence of trust on privacy attitudes is assessed within Chap. 6, and on disclosure within Chap. 5. The influence of prior privacy violations and of demographic factors on privacy attitudes is addressed within the Chap. 6.

3.2.1 Item Selection

Similarly to the LPBA scale, the process starts with generating and selecting the initial set of items for the constructs privacy concern and location privacy knowledge. The process for the constructs is explained in the following.

Privacy Concern (WILL)

Privacy concern was measured as a trait using a construct for the *Desire for Privacy* (DFP) according to Morton [111]. The six-item scale was adopted as is and was used in a series of three online studies to confirm the assumed factor structure. The studies are presented in the Appendices (cf. Appendices A.3, A.5, A.6, and A.7). The assumed factor structure could not be identified; the items not clearly loading on the main factor were removed, as well as those that deteriorated the internal consistency of the construct. The remaining three items, rather than measuring privacy concern, measure *Willingness to Disclose* personal information. Face validity was not assessed for this construct.

Location Privacy Knowledge (KNOW)

Questionnaire items were created regarding the data privacy practices with respect to location data according to the data privacy laws in Germany, where the studies presented in this work are conducted. The aim was to assess the extent to which users are aware of information flows in the context of LBS—such as how companies offering location-based services are required to handle the users' information, and when is the location information collected. Additionally, one item was adopted from a questionnaire proposed by Kraus et al. [93].

In the first version of the questionnaire, technical questions regarding location privacy in two major operating systems, iOS and Android, were presented separately to the users of these systems. A pilot survey with the initial 20 items was conducted ($N = 30$, cf. Appendix A.8). Based on the results from the pilot study, the questions that were specific to any operating system were removed to make the questionnaire more general.

Then, item difficulty indices were calculated as suggested by Moosbrugger and Kelava [110], and Wechsung [172] as follows:

$$P_i = \frac{\sum_{v=1}^{N} x_{vi}}{N * \max(x_i)} * 100 \tag{3.1}$$

where

P_i = difficulty of item

Table 3.7 Results of the EFA and internal consistency of the constructs in background factors

Scale	Variance explained	χ^2	p	Initial N of items	Final N of items	Cronbach's α
Willingness to disclose (WILL)	59.63%	n/a	n/a	6	3	0.808
LBS knowledge (KNOW)	30.07%	n/a	n/a	21	4	0.552

$\sum_{v=1}^{N} x_{vi}$ = sum of the score achieved by N participants on item i

$N * \max(x_i)$ = maximum score achievable by N participants on item i

The item difficulty indices can have a range between 0 and 100. As suggested by Bortz and Döring [30], all items with difficulty indices below 20 were discarded as too difficult, and items above 80 as too easy. Eight items were included in the construct that was deployed in an online study (cf. Sect. 6.2), and in a field study (cf. Sect. 7.1). The same random sample (20%, $N = 34$) of the whole dataset was used to calculate the scale statistics as previously.

For KNOW, the explained variance as well as reliability seem rather low. These issues could be explained by the binary nature of the items, which results in few measurement levels.

3.2.2   Factorial Structure of the Constructs

EFA was calculated for the constructs using the unrotated Maximum Likelihood method [172]. The results for the EFA, as well as for Cronbach's alpha [37] for these constructs are presented in Table 3.7.

After EFA the construct for WILL has only three items remaining, which signifies that conducting a CFA using the fit indices introduced in Sect. 3.1.2 results in a just identified model for the construct. For the construct KNOW, the results from EFA are inadequate. Thus, the validation of the constructs cannot be carried out.

The global model was evaluated using a Maximum Likelihood method with a Promax rotation, with two factors extracted. It should be noted that a Heywood case [91] was identified, suggesting a problem with the factor structure. Also, WILL2 was found to be cross-loading on both the factors. However, the loading on the secondary factor is somewhat smaller than on the primary factor, and could be considered non-problematic. The KMO result suggests an acceptable sampling adequacy; additionally, the Bartlett's test of spherity turned out acceptable.

No constructs could be validated for the background factors, and thus, the subsequent results obtained using such constructs should be interpreted with caution.

3.3   Chapter Summary

This chapter presents a development of a questionnaire to measure location privacy beliefs and attitudes as antecedents to privacy behaviour—named location privacy beliefs and attitudes (LPBA). Furthermore, development of constructs to measure background factors influencing such beliefs and attitudes is presented.

The LPBA scale is based on the taxonomy presented in Sect. 2.2, which is motivated by theory of planned behaviour (TPB) [58] (cf. Sect. 2.1.1) and the reasoned action approach (RAA) by Fishbein and Ajzen [59] (cf. Sect. 2.1.3). The RRA is based on the theory of planned behaviour (TPB), which presents the beliefs and attitudes corresponding to certain behaviour as influencing factors to intention, which is postulated to be a direct antecedent to that behaviour. The aim was to create a measure that reliably represents the dimensions on the second layer of the taxonomy.

The resulting model includes perceived risks (RISK), perceived benefits (BENE), as well as perceived norms (NORM). The final step of the development was validating the LPBA using a small dataset from empirical studies; however, the validation process was not successful. The reason for the unsatisfactory fit indices during the validation process might have been the small sample size used. For validating the questionnaire, more empirical data should be collected. Although not validated, the measure including the three constructs shows a satisfactory model fit. No correlation was found between RISK and BENE. NORM had a small negative correlation with RISK, and a strong positive correlation with BENE, suggesting that users who are more susceptible to peers' influence are somewhat less likely to find LBS risky, and more likely to find them beneficial.

The single-indicator variable privacy calculus (PCAL) could not be included in the final model, and will be assessed within this research using regression analysis. Privacy valuation was not part of the development process, and will, similarly to PCAL, be evaluated in the following parts of this book using regression analysis.

The construct created to measure perceived control (CTRL) did not perform well and was left out of further analysis; more work is needed to address the influence of perceived control on behavioural outcomes in the context of privacy.

Finally, to address the first layer of the taxonomy, a construct was created for measuring Location Privacy Knowledge (KNOW), and a construct for measuring inherent privacy concern as suggested by Morton [111] was evaluated using exploratory as well as confirmatory factor analysis methods. The results suggest that the construct does not form one single factor. To improve the model fit some items were removed, and finally, the remaining items in the construct measure willingness to disclose information (WILL). These measures, as well as those presented earlier in this chapter, will be applied in the following parts of this book.

Chapter 4
Perceived Risks and Benefits in LBS

4.1 Study I: Privacy Perceptions in the Usage of Location-Based Applications

Various benefits can be experienced in the usage of LBSs, including being in contact and organizing meetings with loved ones, receiving discounts, and finding services nearby [124, 158]. These benefits, however, come at the cost of diminished privacy [163]. Many existing works evaluating users' expectations and beliefs with respect to location sharing are from the era before such services were largely popular— thus, such results might be of rather hypothetical nature. Location-based services have become largely commonplace, to the point that they are fully integrated into our daily lives. This was, however, not the case in the beginning of the millennium and the smartphone era, but the services were just starting to become possible concepts to consider. When using a freshly adopted service, at first users are mostly rather conservative in their disclosing behaviour, however, starting to disclose more freely with time when becoming accustomed to the technology [76], and when trust in the service has been established [143]. Based on this assumption, it can be considered questionable whether the early studies on location privacy using smartphone applications reflect the users' actual and current privacy attitudes. Therefore, studying location privacy—and location privacy valuations—can be considered more timely now that the users are well accustomed to the services, and also have perhaps more established comprehension of how much they personally value their location information. Therefore it was seen as a necessary step to update the understanding and assess what the actual users of location-based technologies consider when using such services. Some findings from this qualitative work were used for construction of the LPBA scale (cf. Sect. 3.1.1).

To better understand users' expectations when they use LBSs, an interview study was conducted. In this qualitative study, the focus was in assessing the perceptions of possible benefits, risks, and understandings of what happens with user data.

© Springer Nature Switzerland AG 2020

M. E. Poikela, *Perceived Privacy in Location-Based Mobile System*,
T-Labs Series in Telecommunication Services,
https://doi.org/10.1007/978-3-030-34171-8_4

While various other types of LBSs exist, this work focuses on services on users' personal mobile devices. The following sections present the methodology applied in conducting the study, and discuss the obtained findings.

4.1.1   Method

The study was conducted as an exploratory semi-structured interview during December 2015. The following topics were discussed (cf. Appendix A.1 for the basic interview protocol):

- **Usage of LBS.** What kind of location-based applications are used?
- **Perceived benefits in LBS.** What are the benefits that users expect to receive, or have already experienced?
- **Reasons for not using LBS.**
- **Perceived risks in LBS.** What are the risks that users worry that might be associated with the usage of LBS? Are there some privacy violations that users have already experienced?

4.1.1.1   Data Collection

In total 41 participants (14 female) from different demographic backgrounds were interviewed (for age distribution, cf. Fig. 4.1). Convenience sampling was used, while the aim was to reach a broad demographic and cultural distribution. The interviews were mostly conducted at the respective participant's homes or in cafeterias as per the participant's convenience or via video conferencing using systems such as *Skype*. The interviews were mostly conducted in the participant's native language. Prior to analysis, the interview scripts were translated into English.

A mixture of inductive and deductive approaches was used for analysing the data. Existing literature was used as a basis for the codebook, in particular, expected benefits were based on a qualitative study by Tsai et al. [163]. Two independent reviewers coded the interviews on two consecutive rounds; after the first round the codebook was revised, which was used on the second round for coding all the comments in the transcripts. Finally, all disagreements between the two reviewers were resolved for each case in the transcripts.

4.1.2   Results

In this section, the qualitative findings from the interview study are presented. First, the experienced benefits are discussed, followed by a discussion of risks

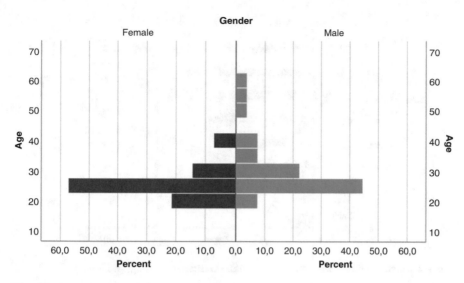

Fig. 4.1   Age and gender distribution of the study

that participants have experienced, or perceive likely. Finally, the mechanisms that participants have for protecting themselves from privacy risks are presented.

4.1.2.1   Perceived Benefits in LBS

The participants mentioned various benefits that they have received from using LBS, and also benefits that they imagine likely, but have not yet experienced themselves. Altogether eight categories of benefits or hypothetical benefits were identified in the transcripts. The benefits, including *navigation, saving time and effort, finding services and location information, social benefits, quantified self, personalized service, safety*, as well as other general benefits are discussed here. These actual and hypothetical benefits, as well as the frequencies at which they appeared in the interviews, are summarized in Fig. 4.2.

Navigation   The most commonly used application type according to the participants' statements were navigation applications. These include maps, navigation aids, as well as applications for public transportation routes and timings. Nearly all participants (90%) stated that they have such applications, and most (71%) also found them beneficial.

Finding Services and Location Information   Almost half of the participants (49%) stated that LBS help them in finding services, including restaurants, stores, or accommodation. Out of these, other information based on location, such as store opening hours or further information regarding a currently visited location, was mentioned by 27% of the participants.

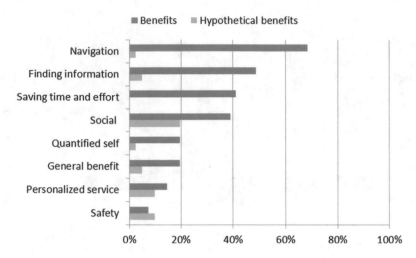

Fig. 4.2 Perceived benefits, as well as hypothetical benefits in the usage of LBS

Saving Time and Effort Forty-one percent of the participants stated that using LBS helps them save time and effort, mostly by requiring less user input and therefore simplifying interaction.

Social Benefits Various social benefits were mentioned by altogether 39% of the participants, including setting up meetings or location-based games with friends. Social recognition, such as letting others know what kind of restaurants or holiday destinations one visits, was mentioned only as a hypothetical benefit. The reason for such benefits to be mentioned only as hypothetical rather than an actual benefit might be that such behaviour could be considered boasting or "showing off", and not socially acceptable. It might be that therefore social recognition is only brought up as something that others do, rather than as a benefit for self. In total, hypothetical social benefits were mentioned by 20% of the participants.

Quantified Self Twenty percent of the participants mentioned that they found it nice to save visited locations in applications that they used, in order to have sort of a diary or mementos, saving memories from past activities and travels. This kind of benefit was mentioned by 15% of the participants. Five percent mentioned the location features of some applications beneficial for sports: biking and running. The benefits include finding routes, as well as tracking the covered distance.

Personalized Service Several participants (15%) also stated that services that have been personalized based on their location are beneficial, including location-based adverts and promotions, or search results that give contextual information based on the location.

Safety Safety features mentioned by the participants included finding lost property, or family members, mainly kids or elderly (7%). Also the possibility for the

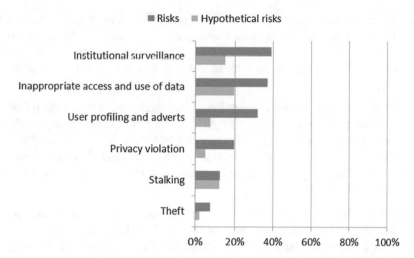

Fig. 4.3 Perceived risks, as well as hypothetical risks in the usage of LBS

government to track citizens as a safety measure was brought up; the possible privacy issues connected with such surveillance were not discussed in this context.

General Benefits Other benefits (20%) that were discussed included personal benefits such as "making one's life better", or "connecting the physical world with the digital one". Also more altruistic benefits were mentioned, such as benefiting the society by generating more data, or helping developers in improving services.

4.1.2.2   Perceived Risks in LBS

Similarly as with benefits, when discussing the possible risks in using LBS, both actual as well as hypothetical risks were identified. The actual risks are risks that the participant mentioned as risks that they are currently concerned about, whereas risks that are considered possible in some circumstances but not likely to happen to oneself are considered hypothetical. Six categories of risks or hypothetical risks were identified, including *institutional surveillance, inappropriate access and use of data, user profiling and adverts, privacy violations, stalking*, and *theft*. These are discussed in this section, and summarized in Fig. 4.3.

Institutional Surveillance Surveillance by the state or police was the most often mentioned risk, brought up as a topic by 42% of the participants. Some commented on such surveillance also as a hypothetical risk, but stated, for example, that they "have nothing to hide", and therefore do not need to worry about being under surveillance.

Inappropriate Data Access and Secondary Use of Data Some concerns were raised regarding personal information of oneself being accessed by individuals that, at the time of the disclosure, were not the intended recipients of the information.

Such concerns were stated by 39% of the participants. Out of these, some were only discussing the risk of the data being visible or accessible by unauthorized individuals, while others also mentioned the risk of the data being possibly used without one's knowledge or consent, in the worst case by individuals with bad intentions.

User Profiling and Adverts Behavioural profiling and advertising based on location information was mentioned by 32% of the participants. Some participants were concerned mainly about receiving targeted adverts, while others stated worries regarding profiles that are built based on behavioural data. Seven percent also discussed behavioural profiling and advertising as a hypothetical risk that did not concern them personally.

Stalking Twelve percent were concerned about becoming victims of stalking as a consequence of location disclosures, or as a result of being a victim of hacking. Further 12% perceived stalking as a hypothetical risk associated with the usage of LBS, including being stalked by one's boss.

Theft Finally, 7% of the participants expressed worries of becoming victims of theft as a consequence of self-disclosures, for example, through social media. Some participants talked about a scenario, where a person with malicious intentions finds out an individual's whereabouts and attacks them based on this information. The perpetrator might even break into the individual's home if, for example, they found out that it is empty through the individual's holiday posts on social media.

Other Privacy Violations In some occasions (20%), participants talked about privacy violations in general without further specifying the type of violation.

4.1.2.3 Understanding Privacy Risks

During the interview, several participants reached for their smartphones to check which applications use location information on their devices. Most of these participants were surprised by the number of different applications that used location.

Half of the participants (49%) revealed that they feel that using LBS involves a trade-off: using such services includes privacy risks, but they accept the risks because they get enough benefits in return. Interestingly though, almost half (44%) stated that they do not think there would be any risks associated with using LBS. Many were of the opinion that they "have nothing to hide", or that only dishonest people would have anything to worry about. Some also stated that when large amounts of data are collected, nothing useful can be done with it, and therefore there are no risks involved.

Several participants talked about being unhappy about the data collection, and experiencing powerlessness with the lack of control. A quarter of the participants (24%) discussed that they do not trust the data privacy practices that companies or the authorities are executing. They also thought that data of them might be unnecessarily collected, saved, or used. Also worries of companies passing information to governmental agencies were expressed.

Trust was expressed somewhat more often than mistrust (29% of the participants). These comments demonstrate trust in companies—many participants trust that in particular big companies treat their customers' private information appropriately. Some participants stated that mishandling users' data would do damage on the company's reputation, and therefore they take care of the users' privacy. Participants also stated that especially the well-known companies have an obligation to treat users' data with care, and that such companies would not sell users' data to third parties.

Some misunderstandings with respect to location privacy were identified during the interview. First, several participants had misunderstandings about some technical details; particularly often the participants thought that GPS was the only way of finding out the location of a device, and by switching the GPS off, the device could no longer be located. Second, some participants seemed to have misunderstandings about what happens to users' data once it has been shared. Some also stated that they do not quite understand how the location sharing and data collection work.

4.1.2.4   Protective Behaviours

Various coping strategies were named by the participants who stated privacy concerns. Such strategies included:

1. **Technical measures**, which were recounted by more than half of the participants (54%). These include switching location services off altogether, or restricting location access either to some applications, or to individuals or groups of individuals through privacy settings.
2. **Avoiding usage** of LBS was mentioned altogether by 39% of the participants; however, only half of them stated explicitly that they avoided usage of location-based applications because of privacy reasons. Apart from privacy, other reasons included saving battery, not seeing benefits in usage of such applications, or annoyance.
3. **Educating oneself** (in total 24%). Participants stated measures such as reading the privacy policies before using a service, or the access rights that an application requests prior to downloading.
4. **No measures taken**. Finally, 22% commented that they do not take any measures to protect their privacy—in some cases even when they feel uncomfortable about their location disclosures.

It could be expected that users who think a lot of risks are associated with using LBSs also are more prone to protect themselves from such risks. Thus, a chi-squared test was conducted to assess the relation of risk perception and taking protective measures. The participants who gave statements that they think there are some risks associated with using LBS also used more protective measures than others ($\chi^2(1, N = 41) = 5.33, p = 0.023$).

4.1.3   Intermediate Discussion and Chapter Summary

Various types of benefits were identified in the usage of LBS, in particular navigation, finding information based on location, saving time and effort because of simplified interaction, and various types of social benefit. On the other hand, also many risks are perceived in the usage, most importantly surveillance, inappropriate access and use of private data, and user profiling, which often is also related with behavioural advertising. The risks included risks caused by individuals, as well as companies and organizations. Some users also seem to have rather optimistic views of what is done with users' data, suggesting, for example, that the (big) companies would not sell users' data because they have a social responsibility to conduct fair data privacy practices.

In this study, the cases of little privacy concern were more often than not connected with the statement that *I have nothing to hide*. This phenomenon has been discussed in the recent privacy literature; Solove states that users who have such a view have a very myopic understanding of what "privacy" means, interpreting it often as "secrecy"—hiding something bad [151]. The prevalence of this view was clearly verified in this study; in some cases there were even statements reporting that there are no privacy risks as long as one has done nothing wrong. These participants have a misconception that privacy equals secrecy; however, privacy should be considered a right, which one can exercise without the need to justify it. Even users claiming the *nothing to hide* argument are likely to expect a certain level of privacy, such as holding a private conversation without being listened to or sharing medical or financial information to a trusted party, expecting the information not to be shared further. Speech interfaces such as Apple's Siri or Amazon's Alexa listen to users' commands, which might be used as a source to show tailored advertisements—a privacy issue that becomes more pronounced when such advertisements are based on private conversations, and in the worst case, displayed on shared devices, giving other users hints of some contents of such private conversations. Similarly, adverts can be based on locational information based on visits that the user had assumed private.

The findings from this study suggest that users who perceive less privacy risks use less methods to protect their privacy. This finding seems rather understandable—when no risks are in sight, also less need for protection is expected. On the other hand, various misunderstandings about information flows were discovered—in some statements even self-reflection about lacking understanding was reported. In many occasions, the statements reflecting misunderstandings also suggested high trust in authorities or companies. Thus, it might be that the users who have limited privacy knowledge also have higher trust and lower risk perception, and therefore use less privacy protection methods. Thus, users could be unknowingly putting themselves at risk if, as a consequence of misunderstandings, they have adopted unsafe privacy behaviours. This can be seen as an indication of that such a situation would be potentially risky if the users are responsible for making sure that their privacy is well protected through limited disclosures and adequate usage of privacy protection mechanisms. This result emphasizes that users should be sufficiently protected against privacy breaches; especially those with limited privacy knowledge, or with lacking motivation to use protection mechanisms should be protected as a default, without explicit user action.

4.1.3.1   Limitations

Some limitations can be identified in the study methodology. Most importantly, there are inherent issues with the self-report method, which might bias the results. First, participants might, even inadvertently, report behaviours that portray them in a better light. This *social desirability bias* is in particular an issue with topics where the participants assume that a certain type of behaviour is desirable; in this study the reported behaviour of reading privacy policies might be prone to such bias. Second, participants may not recall exactly their past behaviours, such as the reasons for installing—or uninstalling—certain applications.

Finally, because convenience sampling was used, the risk for *sampling bias* is rather high in this study. This might lead to low external validity—the results may not be representative of the whole population of smartphone users.

Chapter 5
Location Disclosure and Self-Disclosure

5.1 Study II: Motivation to Disclose Location

To address the question of what motivates users to disclose location using an LBS, a field study with a location-aware mobile participation system called *FlashPoll* was conducted.

Human behaviour is, according to a *self-determination theory* by Deci and Ryan [44], driven by different types of motivations. These motivations can be categorized as intrinsic, extrinsic, and *amotivation*, which is the lack of motivation [44]. When a certain behaviour is intrinsically motivated, it is rewarding in itself, whereas behaviour that is extrinsically motivated is rather an instrument towards a separate, desirable outcome. Amotivation refers to a situation where behaviour does not generate feelings of intrinsic, nor extrinsic motivation, but rather the user can be described as unwilling, uninterested, or unable to engage in an activity. Intrinsic motivation leads to stronger commitment in the behaviour at hand as opposed to extrinsic motivation [142]. This study evaluates what motivates users to disclose location, and whether a relationship between such motivation and disclosure can be identified. Additionally, the influence of feelings of trust towards the data recipient on the disclosure is assessed. Various reasons for disclosure were identified, most importantly wanting to help others. The results from this study also suggest that privacy reasons might be a major factor in unwillingness to install a location-based application.

© Springer Nature Switzerland AG 2020   57
M. E. Poikela, *Perceived Privacy in Location-Based Mobile System*,
T-Labs Series in Telecommunication Services,
https://doi.org/10.1007/978-3-030-34171-8_5

5.1.1   Method

The study included four parts: a pre-study, an online questionnaire, a field study, and a debriefing (cf. Fig. 5.1). Participants for the study were recruited through billboard adverts, as well as through the online participants' database of TU Berlin, *Prometei*. The participants were rewarded with an incentive of 5€ for participation in the online study and the debriefing, and an additional 10€ for participating also in the field study.

The four parts of the study are explained in detail in the following sections:

Pre-study

The pre-study was conducted with 11 expert participants to evaluate the sensitivity of the material to be used within the field study. The material of the pre-study consisted of questions which were used in short questionnaires, or *polls*, within the field study. The participants of the pre-study were presented with each question, and asked to rate them on an end-labelled seven-point Likert scale from "Not at all sensitive" (0) to "Extremely sensitive" (6). The material was then distributed to different polls such that each poll had the same overall sensitivity. Additionally, two somewhat different versions of each poll were created in order to further minimize the impact that the sensitivity of different questions might have on

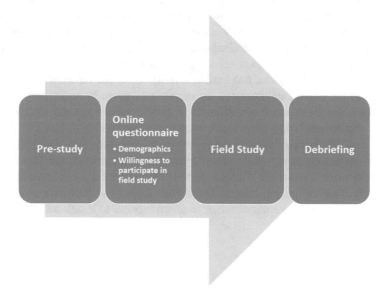

Fig. 5.1 The flow of the study II. The study consisted of four parts: a pre-study, an online questionnaire, a field study, and a debriefing. The aim of the pre-study was to balance the material used in the field study based on sensitivity of questionnaire items. The online questionnaire consisted of a demographic questionnaire and a preparation for the field study. Within the field study, participants received short location-based questionnaires, and clarified the motivation why they would respond to these. The debriefing gave details to the goal of the field study

disclosure; the two versions were distributed between the participants at random such that each participant received one. Some polls required an email address to deliver the participants their monetary incentive. Email address, however, is a single questionnaire item that according to the results from this pre-study is considered very privacy sensitive. To even out the discrepancy in sensitivity levels between all polls, email address was enquired in each one.

The final polls created from these questions did not differ in their sensitivity levels ($Kruskal-Wallis$: $H(15) = 15.0$; $p = 0.451$); therefore it can be assumed that the differences found in disclosure levels were not due to the differing sensitivity levels of the polls (for all polls, see Appendix A.2).

Online Questionnaire
The online questionnaire was conducted using the LimeSurvey platform. It consisted of a demographic questionnaire, and of an introduction to the field study. The participants could then decide to install the mobile participation application on their mobile devices in order to participate in the field study or not to participate, and move straight to the debriefing.

Field Study
The 2-week field study was conducted using the mobile participation tool FlashPoll (cf. Sect. 5.1.1.1). Each of the participants who downloaded the application received altogether eight short questionnaires, or *polls*, during the duration of the field study. The polls were designed such that participation in the various polls gave participants four different types of benefits, and each benefit was present in two polls from different senders, or *poll providers*. The poll providers were added to the experimental design such that each poll was seemingly from a different organization or company to create a higher credibility for the sent polls. What the participants responded in the polls was not analysed, but rather the interest was in whether or not the polls were responded to.

After each interaction with the FlashPoll app the participants were asked to give some details regarding the app usage. This was done by an auxiliary questionnaire app, which opened automatically after each usage of the FlashPoll app. Within this auxiliary questionnaire, the participants were asked why they had used the FlashPoll app, and in the case that they had opened a poll, how much they trusted the company or organization behind that poll—the poll provider.

Debriefing
The debriefing was presented directly after the online study to those participants who chose not to download the FlashPoll app and participate in the field study. For all others, a separate link was provided for the debriefing. It clarified the motivation behind the study and the used deception. The participants who did not download the FlashPoll app to participate in the field study were additionally asked for a reason for non-participation.

5.1.1.1   Apparatus

A mobile participation application *FlashPoll* created in a collaborative, similarly named project within the framework of the European Institute of Technology (EIT) was modified for this study. The main functionality of the application is that short questionnaires, or *polls*, can be established for a certain geographical area known as *geofence*. When a participant enters such a geofence, the participant's device notifies the backend. Subsequently, the backend notifies Google in order to send a push notification to the device that there is a poll available. The devices registered to the FlashPoll service inform the backend about their location occasionally when the device is far away from any active geofence, and with an increasing frequency as the distance to the geofence diminishes. The geofences serve a purpose of reaching users who are physically in a certain geographical area, thus reaching users who are likely the most relevant for citizen participation involving that area.

The application was modified such that the polls within the study were only sent to the study participants, and the normal FlashPoll users would not receive them. Rather than using the geofence functionality in order to reach participants locally, the geofences used in this study were defined to be large enough to cover all participants' physical locations. This was done in order to ensure that all participants receive all polls.

5.1.1.2   Measures

The variables measured in this study are *Trust*, *Poll Type*, and *Reason for Disclosure*; also *Reason for Non-Participation* is assessed. These measures are presented in the following sections:

Trust
Trust toward the poll providers was assessed within a short questionnaire using an auxiliary questionnaire app that would open automatically after each usage of the FlashPoll app. The responses were given on a continuous answer scale with values ranging between [0,1]. The response scale was end-labelled with "very untrustworthy" (0) and "very trustworthy" (1).

Poll Type
Each poll provided one of the following benefits to the user:

1. *Monetary benefit*. By responding to the poll a participant would receive an Amazon voucher of 2€, which was sent to a provided email address.
2. *Indirect personal benefit*. Participants would receive an indirect benefit, such as an improved service.
3. *Altruism*. The benefit received by responding to the poll would be, for example, helping the homeless.
4. *No benefit*. The participant would receive no benefits from responding.

The polls are categorized into four different poll types based on the aforementioned benefits that they provide; each benefit was present in two polls from different poll providers. The different types of benefits were chosen based on the research on motivation by Deci and Ryan in their self-determination theory [44], which states that individuals are motivated through intrinsic as well as extrinsic factors, where intrinsic motivators are expected to lead to more committed behaviours. In this study, the first two types of polls represent the category of extrinsic motivators— a direct personal benefit in the form of a monetary compensation, as well as a less obvious, indirect personal benefit. In the case of the third poll type, named altruism, the behaviour is expected to be mainly extrinsically motivated, however, as the benefits are not as clear as in the first two poll types, responding could also be intrinsically motivated. The fourth poll type represents a possible intrinsic motivators: the mere act of responding to the poll could drive the behaviour. Participants experiencing amotivation could result in non-participation in some or all of the polls.

Reason for Disclosure

After each interaction with the FlashPoll app, an auxiliary questionnaire app was automatically launched in order to enquire why the app was used. When a participant stated that they have responded to a poll, they were asked on a single choice scale why they had participated. The possible options for reasons for participation were:

Because. . .

- . . . I wanted to help others
- . . . I was bored and responding gave something to do
- . . . I got some benefits such as improved service or discounts.
- . . . the topic of the poll was important to me and I wanted to share my opinion or knowledge.
- . . . some people whose choices I respect also use FlashPoll.
- . . . some people expect me to use FlashPoll.
- Other: [please specify.]

Reason for Non-participation

The participants who did not install the FlashPoll application and participate in the field study were asked for a reason for the non-participation. The reason for non-participation was first asked as free text, and additionally through a multiple choice question. To define how much each of the reasons contributed to the non-participation, the participants were asked to what extent they agreed with each reason having been a reason not to participate. The responses were given on a seven-point Likert scale with end points labelled with "Completely disagree" (0) and "Completely agree" (6). The given options were:

Because. . .

- . . . I was out of town.
- . . . I was busy.
- . . . I did not want to commit for 2 weeks.

- ... the workload seemed too high.
- ... the incentive was not high enough.
- ... the task was too technically challenging.
- ... I found that participation would have involved a privacy risk.
- ... I found the field study uninteresting.
- ... I did not feel motivated to participate.
- ... I did not want to install the FlashPoll application on my device.

The responses were normalized between [0,1] prior to analysis.

5.1.1.3   Sample

In total 47 smartphone users (32 female) participated in the online questionnaire (for age distribution within the online questionnaire, cf Fig. 5.2, and within the field study, cf. Fig. 5.3). The education distribution is detailed in Fig. 5.4, and that of occupations in Fig. 5.5.

Thirty-six participants took part in the field study, whereas eleven did not download the FlashPoll app. For the age distribution within the field study, see Fig. 5.3.

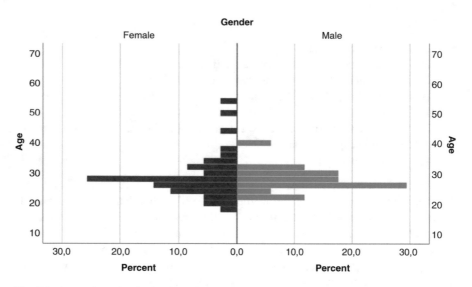

Fig. 5.2 Age and gender distribution of the online questionnaire of Study II

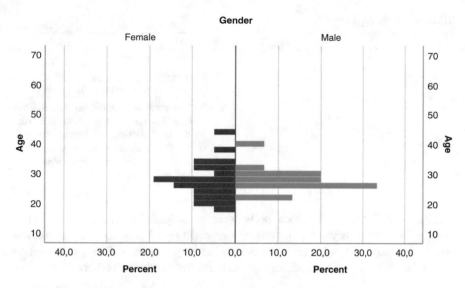

Fig. 5.3 Age and gender distribution of the field study (Study II)

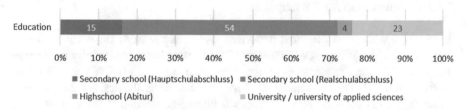

Fig. 5.4 Distribution of educations in Study II

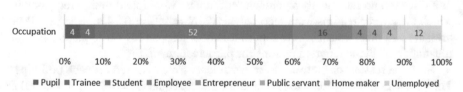

Fig. 5.5 Distribution of occupations in Study II

5.1.2 Results

On average, 77% of the distributed polls were participated in. The results from statistical analysis are presented in this section, and discussed in more detail in Sect. 5.1.3. No impact of age, gender, or education was identified in the results of this study.

First, the influence of different poll types on how frequently polls are participated in was investigated. The most frequently participated poll type was the one named *Altruism* (for more details, cf. Table 5.1). A one-way ANOVA showed statistically

Table 5.1 Ratio of polls
finished by poll type

Poll type	% finished
Altruism	80%
Monetary benefit	79%
No benefit	79%
Indirect personal benefit	69%

The polls that would provide indirect benefit such as improved service were the least participated types. However, no significant differences were found in the participation frequency between any two poll types

significant differences in which polls were participated in most ($F(3704) = 2.95$, $p = 0.032$), however, Bonferroni corrected post-hoc tests did not confirm any statistically significant differences between any two polls. Thus, it cannot be stated that any one poll was participated in more frequently than others.

5.1.2.1 Motivation for Participation

To examine what were the most important motivating reasons for participation, a one-sample Chi-square test was conducted. The test showed statistically significant differences in the frequencies of reported motivations; $\chi^2(6, N = 186) = 90.61$, $p < 0.001$. The most frequently reported reason was the desire to help others. Other frequently reported reasons included finding something to do against boredom, finding the topic important, and receiving a benefit such as an improved service or a discount; no statistically significant differences were found in the frequency in which these three reasons were reported; $\chi^2(2, N = 99) = 0.24$, $p = 0.886$. Peer pressure and other reasons were reported less as reasons for participation. For an illustration of all the reported reasons for participation, cf. Fig. 5.6.

Next, the reasons or motivations to participate were assessed for each poll type. The variances of the reasons to participate were not equal ($F = 47.13$, $p < 0.001$). For the poll type *Monetary benefit*, the main reason for participation was receiving a benefit. The poll type *No benefit* was mainly participated in to get stimulus against boredom. For the remaining two poll types, *Indirect personal benefit* as well as *Altruism*, the most frequently reported reason was the desire to help others. It was the second most frequent reason reported also for the poll type *monetary benefit*. For details on motivation to participate per poll type, cf. Fig. 5.7.

The desire to help others was particularly frequently reported for the poll type *Altruism*, for which it was more than twice as frequently stated as a reason to disclose than any other reason; the difference was also statistically significant ($\chi^2(6) = 44.86$, $p < 0.001$). The polls in this category dealt with homelessness and climate change. There was no statistically significant difference in how frequently desire to help others was stated as a reason for the four poll types ($\chi^2(3) =$

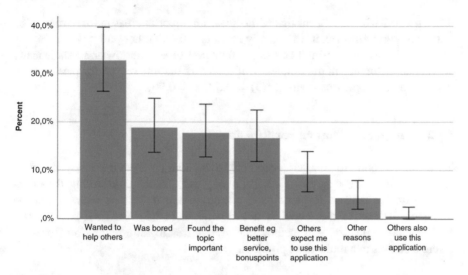

Fig. 5.6 Reported reasons for participation in polls during the field study. The most frequently reported reason was the desire to help others. Finding something to do against boredom, finding the topic important, and receiving a benefit were reported equally frequently with no statistically significant differences between the three

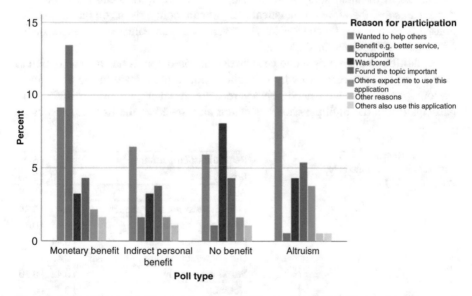

Fig. 5.7 Reason for participation across different poll types. For most poll types, the main reason for participation was the desire to help others. For the poll type monetary benefit, the main reason was receiving a benefit. Getting stimulus against boredom, as well as the importance of the topic, motivated users to participate across all poll types

4.25, $p = 0.236$). This means that while it was an important reason for participation in the poll type *Altruism*, it was equally important for all other poll types.

Similarly, when measured across the four poll types, there was no statistically significant difference in the frequency of reporting that the reason for participation was finding the topic important ($\chi^2(3) = 0.58$, $p = 0.90$).

5.1.2.2 Impact of Trust on Participation

Trust towards poll provider has, according to the findings from this study, no effect on participation frequency ($t = 0.79$, $p = 0.43$). When considering the effect of trust on the reasons to participate, a multinomial regression showed that the influence of trust to the model is not significant ($\chi^2(5) = 7.51$, $p = 0.18$). Multinomial regression is a method that generalizes logistic regression from two discrete classes to a multi-class problem.

5.1.2.3 Non-participation

Finally, the reasons why participants did not install the FlashPoll application and participate in the field study were investigated. There were a limited number of participants and therefore no statistically significant comparisons on the reasons for non-participation can be concluded. However, the mean values of the prominence of the reasons are listed in Table 5.2.

In addition to the multiple choice answers, the participants had also an option to give a free-text answer as to why they chose not to participate in the field study. Various participants who rated privacy reasons as an important factor for non-participation in the multiple choice question also stated in the free-text responses

Table 5.2 Reasons for non-participation in the field study ($N = 11$)

Reason for non-participation	M	SD
Privacy risk	0.70	0.30
Busy	0.55	0.35
Incentive was not high enough	0.45	0.34
Technically too challenging	0.45	0.33
Did not want to install	0.44	0.34
Did not want to commit for 2 weeks	0.43	0.32
Workload is too high	0.42	0.30
Not motivated	0.42	0.32
Study seems uninteresting	0.40	0.24
Out of town	0.32	0.27

The reasons were given on a multiple choice scale with a seven-point answer scale and have been normalized between [0,1]. The main reasons were reported as an expected privacy risk, and being busy that they do not feel comfortable installing various applications—in particular those requiring access to location information—on their devices. Also the worries about losing control over how the data is shared were mentioned in these comments. One participant was particularly clear about the privacy issues by stating: "When I'm on the move, I don't want to always and continuously show and share where I am at the moment. Nowadays there is already enough espionage".

5.1.3   Intermediate Discussion

An empirical study with a location-based mobile participation application called FlashPoll was conducted to study the reasons that users have for disclosing their location information. Various reasons for disclosure were identified, ranging from altruism to getting stimuli against boredom. The most important, or at least the most frequently reported reason was the desire to help others. For the poll type *altruism*, the desire to help others was more than twice as frequently stated as a reason to disclose than any other reason. This finding is easily understandable from the perspective of the poll type, which indeed suggested that others would be helped—in the two polls representing this poll type, these would be the homeless in the city of Berlin, and the climate. The desire to help others was stated equally frequently as a reason to participate for all four poll types, suggesting that the users find altruistic reasons to participate to varying types of polls, even when it is not very obvious that the responding would be beneficial to any other party than the poll providers. Another interpretation of the result is, however, that the participants were conscious about being part of an experiment, and when stating that they wanted to help others, they had particularly the researchers conducting the study in mind.

5.1.3.1   Motivation for Participation

Polls offering indirect personal benefits were most frequently responded to because of a desire to help others, rather than for receiving a better service. The polls were supposedly sent out in order to improve the public transportation network, and mobile network coverage. These polls were designed with such topics based on the assumption that most of the participants can relate to these, and would thus find participation beneficial. However, the stated reasons for participation suggest that the participants did not generally see themselves as beneficiaries of this type of polls, but rather were under the impression that by responding they could help other people. In addition, the topics of these polls were found somewhat important.

One of the polls of the type *no benefit* was supposedly sent by the FlashPoll team, regarding improvements for the FlashPoll app itself. The participants might have considered citizen participation in decision-making an important topic, and therefore found FlashPoll a valuable app that is worth improving. The second poll of this type was supposedly sent by a television company, and the company was also stated as the only beneficiary for the poll. No differences were found between the stated reasons to participate in these two polls of the type *no benefit* ($X(1) = 0.49$, $p = 0.49$); the main reason in both the polls was finding stimulus against boredom.

It is little surprising that what motivated users most to participate in the polls of the type *monetary benefit* was receiving a benefit—a 2€ voucher for a poll that has only 6 easy questions can be considered a rather good deal. The two polls in this category were disguised as a study conducted by a bachelor student, and by a statistical institute. The senders of these polls might explain why the second most prominent reason for responding to this type of poll was the desire to help others, while the other reasons were reported far less. The participants might have felt that these causes are worth supporting.

5.1.3.2 Impact of Trust on Participation

How trust on the recipient impacts disclosure was also addressed in this study. Earlier works have discovered that trust affects disclosure: Bélanger and Carter discuss how increased trust heightens the likelihood of engagement in e-participation [23]. These results were however not repeated in this study. No influence on disclosing behaviour by trust could be identified: there was no relationship between trust towards the poll provider and participation frequency found. Trust also did not influence the reasons to participate when controlled for the poll type. The reason for this discrepancy between the reports from earlier literature and the findings from this study could be in the setting of this study; the participants might have felt obliged to respond to the polls as part of the expectations while participating in a study. To mitigate such a bias raising from an experimental setting, disclosure should be observed in a real-life situation. In this study, there were attempts at avoiding this issue by clarifying that participation in the polls is not mandatory, however, there are no clear indications suggesting whether or not this method was successful.

5.1.3.3 Non-participation

The most commonly stated reason for not participating in the field study and downloading the mobile participation application was privacy concern. One interpretation to this result is that these participants decided to opt out from using the location-based mobile participation app because of uncertainty with how the personal information collected during the app usage would be handled and processed. Another interpretation would be that the participants were unsure about the privacy risks involved in participating in the field study itself; however, the free-text answers provided by the participants support the former view.

The phenomenon of privacy reasons inhibiting adoption is particularly problematic for applications intended to enhance citizen participation when decision-making moves online and to a mobile context. This would create issues of inequality and bias the results as the most privacy-concerned citizens would opt out. Furthermore, the issue of the privacy-concerned possibly having a tendency not to adopt location-based services has inevitably also an implication on privacy research, including the research conducted within this work; the results might be biased and not representative of the whole population as such users are under-represented. Such sampling bias is likely present also in other categories of empirical studies where self-selection is present, as participants are aware that some opinions, attitudes, and possibly behaviours, are measured within the study. However, for privacy studies the issue is more pronounced for two reasons: first, such studies often include rather sensitive questions for probing participants in order to assess their privacy attitudes, and categorize them based on such attitudes [174]. As the primary goal of privacy studies often revolves around finding differences between individuals based on a categorization of privacy attitudes or behaviours, the bias might more severely influence the results, when the most privacy-concerned individuals are under-represented. Second, privacy studies commonly include sensitive aspects, which often means sharing sensitive data with the research team, and potentially also other parties. If the goals of the study are communicated beforehand, the most privacy-concerned individuals are likely to opt out from participating in such experiments in the first place. Thus, privacy studies are more likely affected by self-selection bias than other empirical studies.

The results from this study underline that various reasons motivate users in disclosure using LBS, including altruistic reasons, as well as others as mundane as tackling boredom. The reasons might vary based on the received benefit as well as the recipient of the information. No influence of trust on disclosure was identified in this study, however, privacy concern might be an important inhibitor to adoption of LBS.

5.2   Study III: Location Sharing to Social Relations

This section presents a field study aimed at assessing the effect of social relations on how accurately users share their location information—or whether the recipient of information influences the usage of obfuscation as a privacy protection method. Various recipient types were evaluated, as well as the perceived closeness to each recipient. Also the influence of a reason that the requester of location information states for requesting such information was analysed against the disclosure.

The need for this study was inspired by several issues that were identified with respect to social location sharing. First, most systems provide location sharing restricted only to a binary choice that allows the user either to share or not to share location information. The decision of whether or not the location is disclosed is a complicated process, and involves possibly the privacy calculus, but also various other factors, including the receiver of the disclosed information [21]. The reasons for disclosure might also include reasons based on peers' expectations, enjoyment in using the service, or practical reasons such as receiving benefits [92]. When a user has only such binary choices, all nuances in this decision-making process are lost. As a consequence, when a user discloses location information, it could be assumed that they were comfortable doing so, unless additional information is available. An additional problem with forced binary location disclosure is that in many such situations unnecessarily explicit information of a user is shared, even though less accurate location information would often be sufficient for the purpose of the offered service. To overcome the issue and to appreciate the complex nuances of finding a balance between disclosure and privacy, the users are provided with a range of accuracies in which to share their location information. This method of obfuscation provides the user a privacy control mechanism where the shared accuracy can be chosen on a level that is useful for each situation, rather than sharing an exact location as a default. As an additional location sharing option, the users were also provided with a drop-down menu to share their context.

A second issue faced with while studying location sharing to social relations is how to collect data. Studying location sharing in a real-life situation would possibly provide more data, however in an environment that cannot be controlled. Therefore, a mobile application was created, allowing a simulated environment that was at the same time plausible and controllable.

According to the social penetration theory [12], interpersonal closeness has an influence on the depth and breadth of disclosure between individuals. In relationships in which the individuals do not feel close to each other, the disclosed information is on a rather generic level, and as the relationship deepens, the individuals share increasingly personal information with each other [12]. Therefore, as also previously suggested by Barkhuus [21], the type of relationship between individuals is expected to influence the willingness to share location information. Motivated by these findings, interpersonal closeness between the requester of location information and the participant—the data subject—is evaluated, and its relationship with the location disclosure is examined.

5.2.1 Method

The study was conducted as a field study using the participants' own devices. A messaging application prototype for Android devices was developed to conduct the study; ownership of an Android device was thus a prerequisite for participation. The participants were recruited using adverts on public billboards, as well as through online classified adverts, and an online participation database. The participants received an incentive of 30€ for completing all parts of the experiment.

The study consisted of four parts, as also illustrated in Fig. 5.8:

1. a pre-study within which the most typically shared contextual locations were identified,
2. an introductory meeting within which the study was started and demographic information collected,

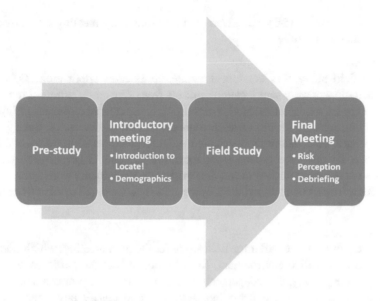

Fig. 5.8 Flow of the study III. The study consisted of four parts. Within a pre-study some typical semantic locations were identified in order to design the mobile application for the field study. Within an introductory session, the participants were introduced to the field study as well as the mobile application. The field study included prompts to disclose location to various recipients. The final meeting included attitude questions as well as a debriefing for the field study

3. a 7-day field study,
4. and a final meeting including a debriefing.

These parts are presented in more detail in the following sections.

Pre-study
The objective of the pre-study was to identify the most typical contextual locations that users report in text form when their physical whereabouts are enquired. This was done in order to create a useful drop-down menu for sharing contextual information within the *Locate!* application, which was used for the main study. The study was conducted using the mobile participation platform FlashPoll (cf. Sect. 5.1.1.1). Fifteen voluntary participants responded to altogether 62 polls during a 4-day period. Convenience sampling was used; no incentive was offered for the participants of this pre-study.

Introductory Meeting
Within the introductory meeting conducted in an office environment, the participants were familiarized with the study procedure, and the Locate! application was installed on their devices. Then, a detailed introduction to the functionalities of Locate! was given. The participants were also asked for names of persons representing different categories of social relations with the participants; the purpose for the data collection was to create test material that is as genuine as possible.

Part of the incentive (5€) was given at the introductory meeting to motivate the participants for the study.

Field Study

A 7-day field study was conducted using the Locate! application. During the study, participants were sent location requests from apparently different requesters, representing six requester categories of varying interpersonal closeness ratings, and with varying reasons for the location requests (for all measures, cf. Sect. 5.2.1.2). The messages were randomly distributed to be received between 8 a.m. and 8 p.m. so as to avoid influence arising from receiving requests at certain times of day, which is not assessed further within this work. The participants were aware that the messages were part of an experiment, however, they were urged to react as genuinely as possible.

Final Meeting

The participants returned after the field study for the final meeting, which was again conducted in an office environment. In order not to bias the participants towards thinking about privacy, the risk perception questionnaire was conducted only during the final meeting. Then, a debriefing was given to clarify any open questions the participants might still have regarding the study and its purpose. Finally, the remaining incentive was given (25€).

5.2.1.1   Apparatus

A messaging application prototype called *Locate!* was developed for the study. The application has a look and feel of a messaging application, with the main functionality of sharing location information (cf. Fig. 5.9). However, the application has limited functionality and cannot be used to share information with other individuals; the software was created exclusively for the purpose of this study.

Upon receiving a location request a user receives a push notification, which opens the Locate! application. The user could specify their location sharing preferences for each location disclosure. To accomplish this, a method of *obfuscation*, or increasing privacy by adding ambiguity, was applied [82]. Obfuscation overcomes the issue of unnecessarily sharing the exact location of a user when a vague location would be sufficient for the purpose. A user could choose the disclosed location accuracy between [25 m, 100 km]. The location accuracies were chosen to be semantically meaningful for the user, such as the accuracy level of a neighbourhood, a block, or the most accurate location. The default accuracy that the user was presented with when the application would open was randomized in order to mitigate the influence of response bias. The accuracy would be chosen from the control buttons in the interface, and additionally for visualization purposes the user has a map within which the currently chosen disclosing accuracy is displayed (cf. Fig. 5.9a). The users were also offered an option to deny location sharing, as well as to share a custom location. Within the option for custom location sharing, the user could define at which accuracy they would like the deceitful location to be shared (cf. Fig. 5.9b).

(a)   (b)

Fig. 5.9 Two screenshots of the Locate! application illustrating some of the available methods for sharing location. The available methods were (**a**) sharing location on a chosen accuracy, which is correspondingly depicted on a map with a red circle; (**b**) sharing a custom location, which can also be done using a chosen accuracy, in which case this is depicted using a blue circle; sharing a semantic location chosen from a drop-down menu, and sharing a comment via a free-text input field. The two latter ones are not illustrated here. The request could also be denied

	English translation	Original German label
Table 5.3 A contextual location could be chosen from a drop-down menu among a list of options	At school	In der Schule
	At home	Zuhause
	At work	Auf der Arbeit
	Commuting	Beim Pendeln
	On the road	Unterwegs
	Somewhere else	Woanders

Additionally, the user could share their contextual location by selecting one from a drop-down menu, and if desired, additionally by writing in free text. The contextual locations for the drop-down menu were created within the pre-study (cf. Sect. 5.2.1), and are listed in Table 5.3.

5.2.1.2   Measures

Several measures were used in this study, collected during the three phases of the study (introductory meeting, field study, and final meeting). Four measures (PUSU, UNAUT, ACCU, and COLL) were collected for risk perception; these have been presented in the Sect. 3. The other measures collected within this study are presented in the following sections.

Requester of Location Information

The participants were asked within the introductory session for names and phone numbers for individuals from different categories of social relations. The categories were as follows:

1. Distant friend
2. Boss
3. Colleague
4. Family member
5. Close friend
6. Partner

The provided information was used to send the participants fabricated messages from the said contacts, asking for the participants' location information.

Interpersonal Closeness

Interpersonal closeness in this study refers to how emotionally close the participants feel with others. Interpersonal closeness was measured within the final meeting on a seven-point scale adapted from Popovic et al. [139], towards each of the individuals listed as a requester of location information. The variable was measured with pen and paper on an illustrative scale that consists of six concentric circles corresponding to the different closeness levels. The circles were labelled with "self" (0), "fully close" (1), "very close" (2), "moderately close" (3), "a little bit close" (4), "neither close nor distant" (5), and "distant" (6) (Fig. 5.10). The statistics for interpersonal closeness are: M = 2.84, SD = 1.52.

Reason for Disclosure

The participants were given various reasons to disclose location information during the field study. A message justifying the location request accompanied each

Fig. 5.10 Interpersonal closeness was measured with pen and paper using a figure consisting of concentric circles labelled from "self" to "distant". The scale is adapted from Popovic et al. [139]

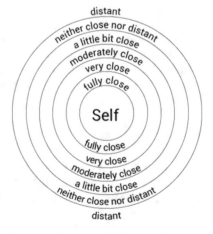

request. The categories for the reasons, as well as an example of an accompanied justification, were as follows:

1. **Positive** "Hi, where are you? Would you like to go for a coffee?"
2. **Neutral** "Hi, where are you?"
3. **Negative** "Hi, there is an issue we need to discuss face to face. Where are you?"

To verify that the categories were indeed perceived as expected, the participants were asked within the final meeting how pleasant they felt when receiving each request. The pleasantness was measured with pen and paper using a seven-point Likert scale end-labelled with "Very unpleasant" (0) and "Very pleasant" (6). A manipulation check revealed that the messages in the *positive* category were perceived as more pleasant than those in the *neutral* category, both of which were perceived as more pleasant than the messages in the *negative* category.

Location Sharing Accuracy
Location sharing accuracy was chosen by the participants within the Locate! interface, and measured on an ordinary scale with eight levels, where items with larger labels denote less accurate location sharing:

1. 25 m
2. 100 m
3. 500 m
4. 1000 m
5. 10,000 m
6. 100,000 m
7. custom location
8. denied

5.2.1.3 Sample

In total 22 individuals participated in the study (for a demographic distribution, cf. Fig. 5.11). The distribution of educations is detailed in Fig. 5.12, and occupations in Fig. 5.13. All participants were active smartphone users. One participant stated technical issues, and was subsequently left out of the further analysis.

5.2.2 Results

The participants received altogether 386 messages asking for their location information, 352 of which were also responded to (91.2%). In 5.5% of the cases that participants responded to, the location information was not shared, which means that the participant chose the option "denied". In this section, the results from the responded polls, as well as from the questionnaire data are analysed. No results were

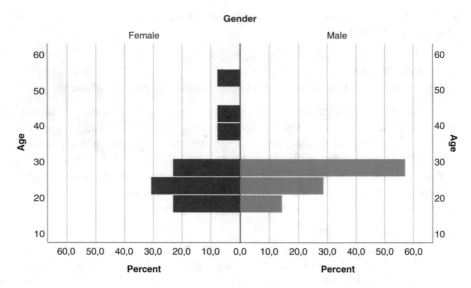

Fig. 5.11 Age and gender distribution of the field study (Study III)

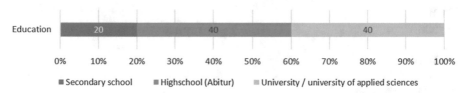

Fig. 5.12 Distribution of educations in Study III

Fig. 5.13 Distribution of occupations in Study III

found based on the age, gender, or level of education of the respondents. The results are discussed further in Sect. 5.2.3.

First, it was examined if all location sharing accuracies were used equally frequently; a one-way chi-square test indicated statistically significant differences in the variances of different location accuracy categories ($\chi^2(7) = 62.63$, $p < 0.001$). The most accurate option was the most commonly chosen option, whereas the least accurate (100 km radius), as well as denying location sharing, were used only scarcely (cf. Fig. 5.14).

A small positive correlation was discovered between location sharing accuracy and the risk perception on the scale ACCU ($r_s = 0.23$, $p < 0.001$). The participants

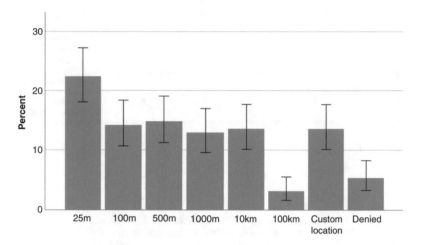

Fig. 5.14 Distribution of location sharing accuracies. The most frequently used sharing accuracy was the most accurate one, whereas the least accurate one, and denying location sharing, were seldom used. The error bars denote 95% confidence intervals

who perceived higher risks measured on this scale shared their location with less accuracy.

Participants shared location more accurately in situations in which the location request was perceived more pleasant, ($r_s = -0.15, p = 0.007$). Additionally, a small negative correlation was found between the closeness score and feeling pleasant about the location request ($r_s = -0.18, p = 0.002$), meaning that the requests from individuals with stronger interpersonal closeness were perceived as more pleasant. The requester role per se did not have an impact on location sharing accuracy ($\chi^2(35) = 34.07, p = 0.513$), but there was a connection between the requester role and closeness ratings ($\chi^2(30) = 289.01, p < 0.001$). Interestingly, users scoring higher on the scale ACCU were more likely to give lower scores on interpersonal closeness, which denotes tighter proximity to other individuals ($r_s = 0.13, p = 0.021$). This indicates a relationship between being concerned about privacy, and feeling close to others.

Semantic locations, or contexts, were shared using a drop-down menu in addition to sharing location coordinates with a chosen accuracy; a semantic location was shared in 59.1% of the responses. A chi-square test revealed differences based on a context in how accurately a location is shared ($\chi^2(28) = 67.96, p < 0.001$, cf. Fig. 5.15). The most accurate location sharing happened most frequently at home. When participants shared their context as "on the road", the most often used sharing accuracies were the most accurate option, and custom location.

A free-text answer was provided in 81.7% of the cases when a location request was responded to. These answers were categorized based on their content; see Table 5.4 for details on the frequency of each category. Two-thirds of the free-text answers described a semantic location, which was often at home, visiting somebody, or on their way somewhere, and out of these, one quarter also suggested to set up

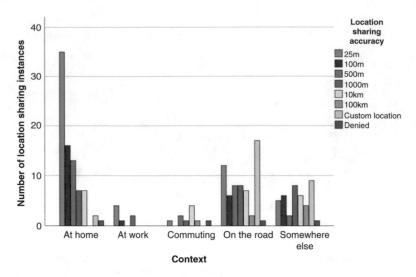

Fig. 5.15 Differences were found in how accurately a location is shared when different types of contextual information is also shared

Table 5.4 Relative frequencies of different types of free-text answers

Content of free-text message	% of responses
Semantic location	67.2
Setting up a meeting	28.0
Excuse for not being able to meet	10.8
Asking for the reason to meet	8.6
Other	2.6

Some responses fall into two or more of the categories. The most typical free-text answer depicted a place where the participant was at a time, mostly to either state that meeting was not possible or in order to set up a meeting

a meeting. The second most frequent type of a free-text response was an attempt to agree on a meeting, and two-thirds of such cases the participant also stated their semantic location. In some cases the free-text field was also used to enquire what reasons the location requester had for having contacted.

5.2.3   Intermediate Discussion

This section presents a study examining location disclosures among social relations. For the study, a prototype of a location sharing application was created, providing the users different methods for sharing location. These methods included sharing a physical location on a chosen accuracy, sharing a custom location, sharing a semantic location from a drop-down menu, and sharing a free-text answer. The location request could also be denied.

Brown et al. presented a device for sharing semantic location and increasing situational awareness between family members [31]. Following along a similar path, the users of the prototype presented in this study could also share their semantic location. The semantic location to be shared could be chosen from a drop-down menu, which had been pre-defined based on results from a pre-study. As a future design suggestion, it would presumably be more meaningful to let the users define the categories for semantic sharing, as big variations between users are likely. Additionally, sharing a semantic location could be offered also as an alternative method of sharing location information, and not only as an additional piece of information as in this study.

5.2.3.1 Interpersonal Closeness

The findings from this study reveal that location requests from individuals that users feel emotionally close to are seen as more pleasant, a finding that does not seem very surprising. There was no direct connection found between the closeness of the participant and location disclosure, however, location was shared more accurately in situations where the location request was perceived as pleasant. The social role of the requester—whether they are colleagues, family members, or acquaintances—does not have a direct influence on the location sharing accuracy. It does, however, have a relationship with how close these individuals are perceived. These results continue the conversation started by Barkhuus [21] on sharing location in social context, in that, whereas the social role of an individual is correlated with how emotionally close one feels with them, the interpersonal closeness in itself can be considered a more meaningful predictor for location disclosures. This finding is also quite understandable, as not all social ties can be measured on similar emotional scales.

5.2.3.2 Privacy Strategies in Disclosing Location

Perception of risks was in this study found to be connected with lower accuracy in location disclosures, suggesting, as proposed in the taxonomy (cf. Sect. 2.2), that risk perception influences disclosing behaviour. This result underlines the importance of minimizing the feelings of discomfort resulting from perception of risks in usage of LBS. The most efficient way to do this is possibly through increasing transparency in how users' data is collected and handled.

Earlier studies have suggested that, rather than being deceitful with disclosure, users sometimes prefer to ignore requests and avoid disclosure altogether [77]. In some situations though, avoiding response could be considered inappropriate, and if the user does not feel comfortable about sharing their actual location, sharing a custom location could suit the purpose. The results from this study suggest that avoiding disclosure, in particular by denying the request, is not a favourable option, and users rather choose an alternative method when they wish to improve their privacy.

5.2.3.3   Disclosing Home Location

Often sharing the context "home" was accompanied with a very accurate location disclosure. This might be surprising in the light of privacy, as users would be expected to see home as a private location. However, considering that the contextual information was given by the participants as additional information for the requester regarding their whereabouts, it can be assumed that when sharing information that one is at home, they have already assessed the situation as not sensitive. Against this explanation it becomes more understandable that then sharing an exact location is not an issue—in many cases the recipient of the information might anyway know where the participant lives.

5.2.3.4   Disclosing Location on the Road

The results suggest that there are two main response categories for location requests when users arc *on the road*. First, a custom location is shared, and second, an exact location is shared. Possible reasons in this case for sharing a custom location could be that a location sharing request arrives in a context or from an individual with whom the participant prefers not to share their current location, and decides to share a more socially desirable, or a less privacy infringing location instead. The shared custom location could also refer to the location where the person was heading to, which might sometimes be a more useful piece of information than the location while in transit. The request could also arrive when a user is expected to be somewhere where they are not, and a custom location provides a way to give a more socially desirable report—or a more useful piece of information by sharing the location where they are heading to.

Sharing the exact location, which was the second most frequent disclosing option accompanied with the semantic location *on the road*, possibly stems from a desire to inform others about one's whereabouts in order to facilitate meeting one another, or to provide the recipient of the location information an accurate estimate of the possible time of arrival.

5.2.3.5   Limitations

Running an experiment with an application that has limited functionalities has some caveats, most importantly that this might jeopardize the credibility of the scenarios. As a consequence, the recorded behaviours could deviate from normal behaviour outside of the experimental setting. During the final meeting some participants gave statements implying that the study setting was plausible, such as commenting the emotional reactions they had when receiving location requests from their superiors or from long lost friends. The finding that location requests were perceived with a varying level of pleasantness based on who the requester of the information was also supports the view that the study succeeded in creating a plausible scenario. Unfortunately, the setting had also its caveats. There were several cases where the participant was in fact accompanied by the person who supposedly sent the location request. These cases were identified from the participants' free-text responses, including statements such as: "I'm right next to you", or "I'm at your place". These kind of cases where it was clear that the hypothetical scenario was not plausible were removed from further analysis.

These results apply to location sharing between individuals, and would possibly look rather different for location disclosures where the recipient of the information is a company rather than an individual. It should be also noted that this study might be influenced by self-selection bias. Additionally, the study was conducted in Berlin, and the results may not necessarily be applicable to the general population.

5.3 Chapter Summary

This chapter focused on disclosure—on the reasons and inhibiting factors of it in two different scenarios. In the first part of the chapter, why users disclose location information was studied within the context of location-based mobile participation. Various reasons were identified motivating users to disclose their location in order to participate in short questionnaires, most commonly the desire to help others. Also reasons such as receiving a benefit, finding the topic important, and being bored were reported as reasons for participation. Privacy concern was identified as a major influencing reason for participants not to install the mobile application and participate in the field study.

In the second half of this chapter, users' location sharing preferences were assessed further within the context of social location sharing. The results from this study indicate that when users have an option to decide on the accuracy in which they disclose location information, most commonly the most accurate option is chosen. Location was shared more accurately in pleasant situations, and by users who perceive less risks in the usage of LBSs.

The next chapter moves from evaluating why users share location to whether the risk–benefit assessment in the context of location privacy can be quantified beyond the observed location disclosure rates.

Chapter 6
Quantifying Location Privacy

6.1 Study IV: Valuation of Location Privacy: A Crowdsourcing Study

As presented in the proposed taxonomy in Sect. 2.2, LPV is assumed to be a measurable outcome of the privacy calculus, which is the assessment of the perceived risks and benefits. Thus, LPV can be considered an operationalization of privacy calculus, providing a measurable factor. In this chapter, the core question is whether LPV can be considered such a variable that can be empirically measured and used to quantify privacy calculus, and to predict location privacy behaviours. In this work, monetary valuation is deployed as a method for this quantification. Using money as a measuring instrument offers a fine-grained method for estimating differences in privacy behaviours, such as how willingly users share their location.

How much users value their privacy has been assessed in previous works. Users have been found to perceive the received benefits of information exchange so remarkable that they are willing to give away their private data for free [141]. In such a case it is, however, debatable if the information is given away for free, or if the received benefit would be an adequate compensation. According to Tsai et al. [162], when offered a more privacy aware option for the information exchange, users are willing to pay a premium for the added privacy. This finding implies that users can attribute a monetary value to their privacy.

Location privacy valuation has been studied earlier by Danezis et al. [40], who used auction theory to assess how much users would like to be paid to share location information for a certain period of time. The average value stated was 38€, which increased to 45.60€ when participants were informed that the location information would also be used for commercial interests. This finding suggests that the location valuation is likely to reach higher values in privacy infringing situations, such as when advertisers are involved. The findings were later confirmed by Cvrcek et al. [39], who conducted a larger-scale study involving participants from various

© Springer Nature Switzerland AG 2020   83
M. E. Poikela, *Perceived Privacy in Location-Based Mobile System*,
T-Labs Series in Telecommunication Services,
https://doi.org/10.1007/978-3-030-34171-8_6

nationalities. A study by Barak et al. [19] with Android users discovered that users give different values to different contexts, with the median value required to be paid for sharing the location of home being the highest, and location information while in transit the lowest.

This section presents a study addressing location sharing in-situ using a scientifically developed crowdsourcing platform *Crowdee*,[1] which is designed in particular for mobile use. The aim was to assess mobile users' location privacy valuation in various scenarios. The findings from the study suggest a positive relationship between the amount of money paid and the willingness to disclose location information (cf. [137]). Also the scenario in which the location is shared seems to influence disclosure. The amount paid influenced disclosure mainly in scenarios where an advertiser was present. The disclosure rates were overall rather high, even in scenarios where there were no obvious benefits from disclosing. However, in a scenario where an untrusted advertiser was involved, the sharing rates dropped. This suggests that trust towards the recipient of location information has an impact on willingness to disclose.

6.1.1   Method

This study consisted of three parts: a field study with the mobile crowdsourcing platform Crowdee with German participants, and two follow-up questionnaires, which were also administered using the Crowdee platform (cf. Fig. 6.1). Using the crowdsourcing method provides a possibility to conduct the study in-situ rather than hypothetically. An additional benefit in using the platform is that the crowd workers are already accustomed to conducting small tasks for monetary rewards.

Field Study
During the field study the users of the Crowdee platform, or *workers*, were offered *jobs* with a base payment of 0.10€. Jobs are short tasks that the workers choose within the platform, for which they earn money based on the duration of the job and required additional qualifications. The base payment was offered to provide compensation for the workers for taking the job irrespective of whether or not they disclosed location. The workers could earn an additional bonus for sharing their location information. The amount of the bonus, randomized between [0€, 0.50€], was stated within the job before the worker chooses whether or not to disclose location. No other questions or tasks were included; the workers could either disclose location and earn the base payment and additionally the bonus, or not disclose, and earn only the base payment.

When a worker had accepted a job, they were presented with one of four possible scenarios, chosen at random. The scenarios represented different potential data

---

[1]https://www.crowdee.de/en/.

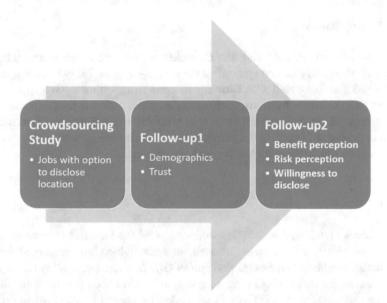

Fig. 6.1 Study IV was a crowdsourcing study within which crowd workers selected jobs, and could disclose location against a small monetary compensation. Demographic and attitudinal questions were divided between two follow-up questionnaires

recipients and subsequent uses of data, including sharing presumably only Crowdee, sharing with two different advertisers, and a social sharing scenario, in which the location would be shared with other Crowdee users.

A worker who was deemed eligible for the job could take the job for up to ten times. A buffer time of 120 min was enforced between the jobs to avoid multiple jobs being taken in the same location.

Follow-up Questionnaires

The demographic as well as attitudinal questionnaires were delivered to the crowd workers who had taken at least one job during the field study. The questionnaires were separated from the field study to keep the jobs restricted to the location sharing scenario, and divided into two parts to keep each at a reasonable length for a mobile device.

The first follow-up questionnaire was published after the field study was over. It included a demographic questionnaire, and altogether sixteen questions regarding perceived trustworthiness of the alleged data recipients in each of the four scenarios: Crowdee, the two advertisers, and other Crowdee users (cf. Sect. 6.1.1.2 for details).

The second follow-up questionnaire was published shortly after the first one, and included questions covering risk perception, benefit perception, as well as willingness to disclose.

6.1.1.1   Apparatus

The study was conducted using the Crowdee crowdsourcing platform, which is created by a similarly named spin-off of the Quality and Usability Lab. A webview was created and integrated with Crowdee to open from within a job to provide the required functionalities for sharing a location, as these were not readily available on the platform. The webview included a map with the user's actual location marked, a short explanation regarding the purpose of data collection as well as the provided bonus, and finally, buttons for sharing and not sharing.

6.1.1.2   Measures

The variables considered within this study included the *location sharing scenario*, *location sharing rate*, *payment*, as well as *trust*, which are presented here in more detail. Additionally, *benefit perception (BENE)*, *risk perception (RISK)*, and *willingness to disclose (WILL)* were assessed, these have been presented in Chap. 3. Trapping questions were used as recommended by Naderi et al. [114] in order to ensure the high quality of responses.

Location Sharing Scenario
Four different scenarios were created, and presented randomly to participants taking the job. The scenarios, and explanations given for why the data was collected, were as follows:

1. **Sharing location with the crowdsourcing platform.** "Share your location with us. The data will be used by us for customer behaviour analytic purposes".
2. **Sharing location with a trusted advertiser.** "Share your location with us. The data will be used for customer behaviour analytic purposes by a third party".
3. **Sharing location with an untrusted advertiser.** "Share your location with us. The data will be used for customer behaviour analytic purposes by a third party".
4. **Sharing location in a social setting with other users of the crowdsourcing platform.** "Share your location with us. The data will be used for customer behaviour analytic purposes by a third party".

In each scenario, a webview opened with a map, showing the participant's current location. Additionally, the explanation for data collection purposes was provided, as well as a clarification that for sharing the current location, the participant would be paid a certain amount of bonus (cf. Fig. 6.2).

In Scenario 1, no other cues were provided. This scenario can be considered as a baseline for measuring trust, because the users of this platform could be expected to have an established level of trust towards the platform.

The Scenarios 2 and 3 additionally had a fabricated advert by a trusted advertiser (Scenario 2), or by an untrusted advertiser (Scenario 3). The adverts linked to the respective companies' webpages; the companies were not informed about the study, nor were they involved in any other way in the study. As the trusted advertiser, a

(a)   (b)

Fig. 6.2 Two screenshots of the webview presented within the Crowdee platform illustrating some of the four scenarios, including (**a**) sharing location with a trusted advertiser and (**b**) sharing location in a social setting with other Crowdee workers. The other scenarios were sharing location with an untrusted advertiser and sharing location with Crowdee. The screenshots additionally illustrate that the workers were, prior to sharing, provided information regarding the bonus they would be paid for the shared information in that scenario

company ranking among the top five in a study assessing the impact that 127 German and international companies have on general wellbeing, was chosen. The chosen untrusted advertiser was a company ranking within the bottom five in the same study. Both the companies are well known in Germany to the extent that familiarity with both can be assumed. It should be noted that these two scenarios had identical wording for the clarification of the data collection purposes in order to avoid any differences in attitudes or behaviours based on these wordings; the only trigger difference between these two scenarios is the type of the advertiser. Furthermore, it was not explicitly stated that this advertiser would have access to the user's location.

In Scenario 4, fabricated profile cards of other crowd workers using the same platform were shown on the map around the participant's location to give a feeling that there are some actual users with whom the location would be shared. No actual information from other crowd workers were used to produce such profile cards. Some of the produced profile cards had profile pictures on them for increased credibility; the pictures were collected from open source databases.

Location Sharing Rate

Location sharing is measured as a binary variable denoting whether or not the location sharing task within a job was accepted by pressing the "Share" button within the webview. The sharing task was denied if "Do not share" button was selected, and also if the participant returned from the webview by pressing the back button.

Payment

The participants would receive a base payment for accepting a job, and an additional bonus for accepting a location sharing task within the job. The amount of bonus for each location sharing task was between 0€ and 0.50€. A uniform distribution of bonus payments with 0.1€ intervals was randomly distributed across all jobs. The base payment was not considered in the analysis.

Trust

Perceived trustworthiness of each receiver of location information presented in the four scenarios was measured using four items. The items were measured on fully labelled bidirectional seven-point Likert scale with labels such as "Very trustworthy" (0), "Somewhat trustworthy" (1), "A little bit trustworthy" (2), "Neither nor" (3), "A little bit untrustworthy" (4), "Somewhat untrustworthy" (5), and "Very untrustworthy" (6). A score for trust towards the data recipient in each scenario was measured by taking a mean value over the four items. Additionally, overall trust was measured by taking a mean value over the scores from each scenario (See Appendix A.10 for all items, and Table 6.1 for internal consistencies).

To assess whether or not the participants perceived the trustworthiness of the advertisers as expected—higher for trusted than for untrusted advertiser—a one-way ANOVA was conducted. The results suggest that there are differences between the trustworthiness scores of the four data recipients ($F(3, 424) = 162.46$, $p < 0.001$). The trusted advertiser was perceived as equally trustworthy as the crowdsourcing platform, and both were perceived as more trustworthy than untrusted advertiser, or other users of the crowdsourcing platform (cf. Fig. 6.3).

6.1.1.3   Sample

The participants of the study were existing German-speaking crowd workers of the Crowdee platform within Germany. The job was offered only to workers who had taken a German language job and thus proven eligibility for the task. Altogether 190 unique crowd workers accepted the job, each of whom could repeat it up to ten times. From the 190 workers that participated in the first part of the study, 109 took part in the first follow-up questionnaire, and 116 in the second follow-

Table 6.1 Measures for trust for each data recipient according to the four scenarios, and the Cronbach's α of each measure

Measure	No. of items	Cronbach's α
Trustworthiness of crowdsourcing platform	4	0.734
Trustworthiness of a trusted advertiser	4	0.781
Trustworthiness of an untrusted advertiser	4	0.768
Trustworthiness of other users of the crowdsourcing platform	4	0.750
Overall trust	16	0.857

Each measure had an acceptable internal consistency measured by Cronbach's α

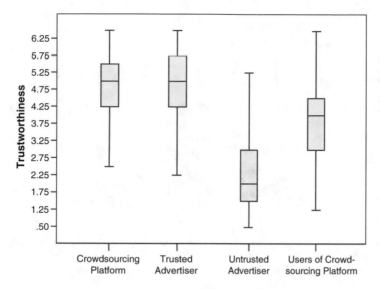

Fig. 6.3 Differences were found in how trustworthy the four data recipients, each representing a different experimental scenario, were perceived. Trusted advertiser was perceived as more trustworthy than untrusted advertiser. The crowdsourcing platform was perceived equally trustworthy as trusted advertiser, and more trustworthy than untrusted advertiser, or other users of the crowdsourcing platform

up questionnaire. After removing low quality responses based on the trapping questions, 105 responses were accepted for the first follow-up questionnaire, and 107 for the second follow-up questionnaire. The remaining participants were disregarded for any further analysis.

The crowd workers who participated in the study were mainly young adults ($M = 28.76$ yrs, $SD = 8.83$), with 60% male workers. The age and gender distribution of the study is presented in Fig. 6.4. One-third of the workers had a university degree (cf. Fig. 6.5), and 46% were students or pupils (cf. Fig. 6.6). A total of 35% stated working, or having in the past worked, in a field related to IT. Note that this data represents the workers who participated in the first follow-up questionnaire. It can be assumed, however, that it is a good representation of the whole sample of the study.

6.1.2 Results

In total 1064 jobs were taken. Seventy-two of these were not carried out completely, which means that the crowd worker left the job without choosing to share or not to share location. Of these 72 cases, 58 had problems with location setting, and therefore the map could not be displayed properly; the remaining 14 cases are

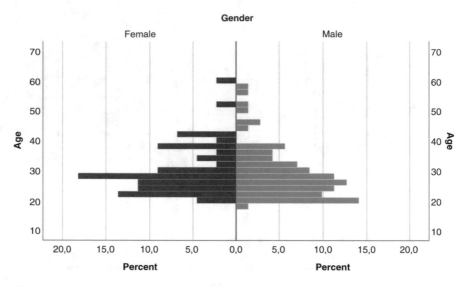

Fig. 6.4   Age and gender distribution of the Study IV

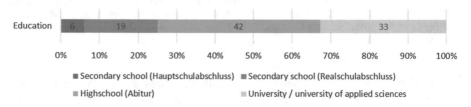

Fig. 6.5   Distribution of educations of crowd workers in Study IV

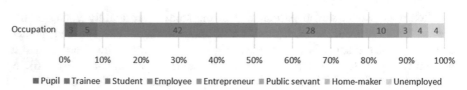

Fig. 6.6   Distribution of occupations of crowd workers in Study IV

handled in the further analysis as "not shared". All cases that did not pass the trapping questions in the follow-up questionnaires were disregarded from analysis. Additionally, the analysis was restricted to participants who shared their location at least once in order to concentrate on the reasons why location sharing did happen. The following analysis is done based on the remaining 435 jobs that were carried out. Location was shared in 84% of these jobs taken. The statistical analysis carried out is presented in this section, and discussed in further details in Sect. 6.1.3.

6.1.2.1   Location Sharing and Payment

Considering the acceptance rate of the location sharing task per payment level, a statistically significant effect on sharing by payment was identified—in the cases in which location was shared, the payment was significantly higher ($t(433) = -4.87, p < 0.001$). Location sharing probability was then modelled as a binary logistic regression, given by

$$P(location \quad sharing|x) = \frac{1}{1 + e^{-(\Theta * x + b)}} \tag{6.1}$$

where

$\mathbf{x} = (x_1, x_2, \ldots, x_n)^t$ the constituent variables and
$\boldsymbol{\theta} = (\theta_1, \theta_2, \ldots, \theta_n), b$ corresponding model parameters.

The two-class problem of whether or not a location was shared as a function of payment (x_1) was estimated using the Eq. (6.1). The following parameters were obtained for the model: $\theta_1 = 0.70, b = 4.66$. This result suggests that the location sharing rate is strongly related to the payment, however, the model explains less than 10% of location sharing behaviour (*Nagelkerke $R^2 = 0.09$*).

It can be assumed that acceptance of location sharing task follows a logarithmic model, where a higher payment yields higher acceptance, and finally plateaus after a certain threshold. This assumption is next applied by taking the percentage of accepted location sharing tasks per payment level, and fitting this to the logarithmic model. This gives a prediction for the percentage of workers who share their location for a certain payment (cf. Fig. 6.7). The model explains a significant proportion of location sharing behaviour ($R^2 = 0.44, p < 0.001$).

6.1.2.2   Location Sharing per Scenario

The influence of the scenario on location sharing was assessed; a Kruskal–Wallis H test showed a significant difference between the sharing rates in the four scenarios ($\chi^2(3) = 15.22, p = 0.002$). Bonferroni-corrected pair-wise Mann–Whitney U-tests showed that location sharing tasks were accepted less frequently when an untrusted advertiser was involved, than in other scenarios (cf. Table 6.2).

To further investigate the influence of each scenario, the data was divided based on the four scenarios, and the probability of location sharing per payment level was assessed for each scenario. All four cases were analysed separately using the binary logistic model. For Scenarios 1 and 4 the model turned out not to be statistically significant. For Scenarios 2 and 3, the parameters for payment (x_1) are presented in Table 6.3. The logistic regression model for the two scenarios involving advertisers, as well as location sharing with the whole dataset, are illustrated in Fig. 6.8. The results suggest that the probability in which location is shared depends on the payment mainly in the case that an advertiser is involved; the models were not

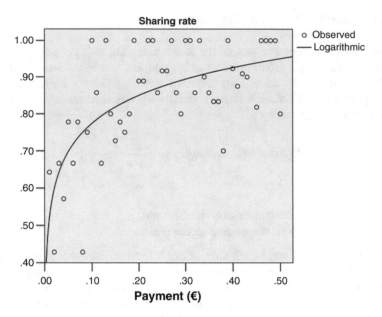

Fig. 6.7 Each point is a certain amount of payment from 0€ to 0.50€—not a sharing rate of individual participants

Table 6.2 Results of Mann–Whitney U-tests with a Bonferroni-corrected alpha level $\alpha = 0.0083$

	Scenario 2	Scenario 3	Scenario 4
Scenario 1	$\chi^2 = 973.0$, $p = 0.851$	$\chi^2 = 669.5$, $p = 0.004$	$\chi^2 = 984.0$, $p = 0.948$
Scenario 2		$\chi^2 = 0.692$, $p = 0.005$	$\chi^2 = 996.5$, $p = 0.862$
Scenario 3			$\chi^2 = 681.0$, $p = 0.004$

The differences between Scenario 3 and other scenarios were statistically significant. All other post-hoc tests turned out not to be statistically significant on the new alpha level

Table 6.3 Parameters of the binary logistic model estimating probability of sharing location as a function of payment in the scenarios of sharing location with a trusted advertiser (Scenario 2), sharing location with an untrusted advertiser (Scenario 3), as well as for the whole dataset including the four scenarios

	θ_1	b	R^2
Scenario 2	9.31*	0.32	0.23
Scenario 3	3.99*	0.13	0.08
Total (4 scenarios)	4.66*	0.70	0.09

For location sharing in Scenario 2, 23% of the variance could be explained by the model, whereas 8% was explained by the model for the Scenario 3, and 9% using the whole dataset

Fig. 6.8 Logistic regression model, giving a probability that a location is shared as a function of payment, in the scenarios of sharing location with a trusted advertiser (Scenario 2), and sharing location with an untrusted advertiser (Scenario 3). "Total" refers to the combined data, including all four scenarios

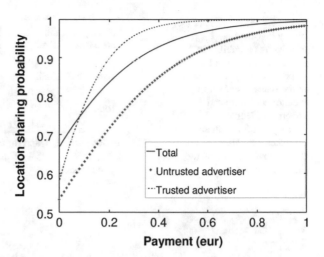

significant in the other two scenarios. Furthermore, a premium needs to be paid for sharing location with an untrusted advertiser in order to reach the same location sharing probability as with a trusted advertiser.

6.1.2.3 Location Sharing and Trust

Next, the influence of trust was assessed, as it is assumed that the differences between the two advertisers are a consequence of differing trust levels.

First, using the whole dataset, the effect of trust was assessed with a t-test, which showed that when location is shared, overall trust is higher ($t(52) = -2.05, p = 0.046$). Thus, trust seems to be connected with more frequent location disclosures. Logistic function was then used to estimate the probability of a location being shared as a function of payment. The model turns out to be significant; the model parameters are as follows: $\theta_1 = 0.71, b = -0.23$. However, the variance explained by the model is very modest, below 6% ($Nagelkerke\ R^2 = 0.057$). Therefore it can be noted that trust and location sharing probability are connected, but trust does not explain the majority of location sharing.

Next, trust was added as a parameter to the binary logistic model, estimating the probability of location sharing as a function of payment and trust in each of the four scenarios. For the Scenarios 1 and 4, the model did not turn out to be significant. The parameters for payment (x_1) and trust (x_2) are listed in Table 6.4 for these two scenarios with advertisers, as well as for the total dataset, in which case the variable (x_2) denotes overall trust. The influence of trust turns out not to be significant in Scenario 2, suggesting that location sharing probability is in the presence of a trusted advertiser mainly influenced by payment, as illustrated in Fig. 6.8. The location sharing probability as a function of trust and payment in Scenario 3, and for the total dataset, is illustrated in Fig. 6.9.

Table 6.4 Parameters of the binary logistic model estimating probability of sharing location as a function of payment (x_1) and trust (x_2) in the scenarios of sharing location with a trusted advertiser (Scenario 2), sharing location with an untrusted advertiser (Scenario 3), as well as for the total dataset

	θ_1	θ_2	b	R^2
Scenario 2	9.60	0.48	−1.97	0.27
	$p = 0.003$	$p = 0.146$		
Scenario 3	3.88	0.53	−1.02	0.15
	$p = 0.026$	$p = 0.018$		
Total	30.82	1.19	−5.03	0.45
	$p = 0.021$	$p = 0.025$		

In Scenarios 2, 3, and for the whole dataset, respectively, 27%, 15%, and 45% of the variance could be explained by the model

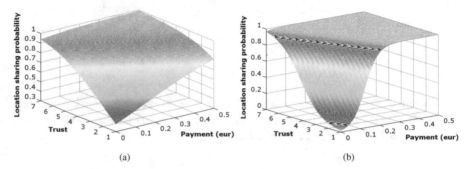

(a)   (b)

Fig. 6.9 Logistic regression as a function of payment and trust in the Scenario 3, location sharing with an untrusted advertiser (**a**), and when considering the whole dataset (**b**). Both variables contribute to the model significantly. In the Scenario 3 (**a**), location sharing probability for a certain level of payment is lower than for the whole dataset. When trust is high, sharing probability reaches rather high levels even in the Scenario 3 (**a**), however, trust was generally quite low towards the advertiser involved in this scenario. In the Scenario 2, location sharing with a trusted advertiser, trust did not significantly contribute to the model, and thus this scenario is not depicted here

6.1.2.4 Location Sharing and Privacy Attitudes

To assess how privacy attitudes affect location sharing in this context, the influence of perceived risks and benefits together with payment on location sharing were analysed. This gives a three-variable model, where variables payment (x_1), perceived risks (x_2), and perceived benefits (x_3) get the corresponding parameter values $\theta_1 = 4.72$, $\theta_2 = -0.38$, $\theta_3 = 0.31$, and $b = 1.05$. Having included these additional variables, the variance explained by the model is 14% ($Nagelkerke\ R^2 = 0.14$), and the model classifies correctly 84.8% of the cases.

The influence of privacy concern as *willingness to disclose* on location sharing was assessed using an independent samples t-test. The results show that privacy concern is connected with sharing a location less frequently; independent samples t-test showed a difference in privacy concern between the cases where the location was shared and those where it was not shared, with concern being higher in cases where location was not shared ($t(52) = 2.193$, $p = 0.033$).

Finally, location sharing was modelled as a function of payment (x_1) and willingness to disclose (x_2). The obtained parameter values were $\theta_1 = 24.24$, $\theta_2 = 0.97$, and $b = 3.89$. The model could explain nearly half of the variance in the sharing behaviour ($Nagelkerke\ R^2 = 0.45$), and classify correctly 82% of the cases. The result suggests that privacy concern and payment together have a strong influence on location sharing behaviour.

6.1.3 Intermediate Discussion

A crowdsourcing study was conducted to evaluate how payment influences location sharing. The aim of the study was to find out if privacy behaviour can be predicted using the method of monetary quantification. This study additionally assessed how disclosing a location varies in scenarios where different potential data recipients are involved. Also the influence of trust in such scenarios was evaluated. Four scenarios were presented to the participants within a crowdsourcing task. These scenarios involved staged situations where the worker was informed what the shared information would be used for (e.g. for customer behaviour analysis by third parties). The results of the study are discussed in this section.

6.1.3.1 Impact of Payment

The amount of bonus paid for sharing a location has a significant impact on acceptance rates of the location sharing task. The location sharing rate per payment level seems to follow a logarithmic model, in which small increments in paid bonus increase the acceptance rate rather strongly. A majority of the users seem to be disclosing location even for very low levels of compensation. The probability that a location is shared is rather high even in the case where no payment was provided—and the user received no apparent benefits for the location disclosure. This highlights the interesting phenomenon of location sharing for apparently no benefit, as also discussed in Chap. 5. Earlier in this work some of the explanations for this kind of behaviour included fighting boredom by having something to do, and helping others—such as the researchers conducting a study. In this case, however, sharing and not sharing location had equal workloads: the crowd worker had to either choose to share location, or choose not to share location; therefore finding something to do against boredom does not seem like a plausible explanation for the sharing behaviour in this context. Also, the crowd workers did not know they were part of a study when accepting the job, but were accomplishing the same kind of work that they, as crowd workers, were already accustomed with. Thus, if they did share location in order to help others, it can be assumed that they did so in order to help the assumed data recipient in each scenario, including the used crowdsourcing platform, or third parties. In the case of sharing with the crowdsourcing platform, the crowd workers might have expected that by disclosing location, they might be advancing research by helping the platform.

6.1.3.2   Impact of Scenario

How likely location sharing is depends on the scenario in which it is requested: compared to any other scenario, sharing was less likely in a scenario in which an untrusted advertiser was involved. Statistically significant differences were found in the trustworthiness of the different instances involved in the four scenarios. The crowdsourcing platform and an advertiser expected to be "trusted" were rated as more trustworthy than other crowd workers of the same platform, and an advertiser that was expected to be "untrusted". Also, the other crowd workers were rated as more trustworthy than the untrusted advertiser.

When assessing location sharing probability per scenario as a function of payment, the model was significant in the scenario where the data recipient was a trusted advertiser, which had the same trust rating as the crowdsourcing platform. However, the model was not significant in the scenario where the crowdsourcing platform was the assumed data recipient. This result undoubtedly suggests that when advertisers are involved, the payment has an influence on the location sharing probability. Two possible explanations as to why the location sharing probability could not be modelled as a function of payment in the scenario with the crowdsourcing platform could be identified: first, it is possible that the relationship between payment and location sharing is indeed different when a third party receives the information, rather than when the data is shared only with the service provider. Second possibility is that the participants of this study were not sure what the reasons for data disclosure were, and what would the data be subsequently used for. Then, the insecurity has possibly been present in both the scenarios (sharing with the crowdsourcing platform, and sharing with a trusted advertiser); however, only in the scenario with an advertiser was the behaviour systematic, and in the scenario involving only the crowdsourcing platform, the interplay of trust and payment was not apparent.

The location sharing model as a function of payment turned out not to be significant also in the scenario where other users of the crowdsourcing platform were the assumed data recipients, which could suggest that in this scenario the subsequent data use was not clear to the participants.

6.1.3.3   Impact of Trust

Trust was found to influence the probability of location sharing, however, not sufficiently to explain it adequately.

The interplay of trust and payment was assessed. Similar results were obtained estimating the location sharing probability on a binary logistic model as a function of trust and payment—as was the case with estimating location sharing as a function of payment, the model was again not significant in the scenarios involving the crowdsourcing platform, or other users of the crowdsourcing platform. Additionally, trust turned out not to contribute to the model in the scenario with a trusted advertiser, suggesting that in that case the sharing probability is mainly influenced by payment. In the scenario involving an untrusted advertiser, both trust and payment contributed to the model significantly. This result suggests that location sharing probability is influenced by the offered monetary compensation when advertisers are present, and when the advertiser is not perceived as trustworthy, also by trust.

In the Scenario 3 (involving an untrusted advertiser), the sharing probability as a function of payment increases following a more moderate curve than in the Scenario 2 (involving a trusted advertiser). This finding suggests that when an untrusted advertiser is involved, a premium is required to reach the same location sharing probability as otherwise, even comparing with a scenario with an advertiser that is more trusted. When no payment is provided for location sharing, the location sharing probability is still rather high, at more than 50% even for the scenario involving untrusted advertiser. In other scenarios, such as that with a trusted advertiser, in the situations where no payment is offered, the sharing probability is even higher.

6.1.3.4   Impact of Privacy Attitudes

Also the impact of privacy attitudes—that of perceived risks, benefits, as well as privacy concern—on location disclosure as location sharing task acceptance rate was assessed. The results of this study indicate that perceived benefits and risks together with payment predict location sharing somewhat better than payment alone, however, payment has a greater impact on location sharing that either perceived benefits or perceived risks.

Privacy concern was found to influence location disclosure; and privacy concern together with payment provide a better prediction of location sharing than payment alone. Thus, the results indicate that perceived risks, benefits, and privacy concern have an impact on location sharing, however, the impact of payment seems to be more prominent. Additionally, trust influences location sharing in certain situations, in particular when advertisers are present.

6.1.3.5   Limitations

Using a crowdsourcing platform enabled studying the impact a payment has on a location sharing task in a realistic scenario, as the crowd workers are accustomed to doing small tasks and being paid for them. The drawback of conducting the study using a crowdsourcing platform with participants who were its existing users is that the results might be biased by the sample. It is possible that crowd workers are more inclined than average to have a desire to help by participating in research. This could offer one explanation for the disclosure rates being rather high.

Another limitation of this study is that for properly assessing the value of location privacy, the users should be knowledgeable of the use, and possible secondary use of the data. This problem is addressed in the following section, which also addresses the influence of location privacy knowledge on privacy valuation.

6.2   Study V: Valuation of Location Privacy: An Online Study

To assess the impact of other factors on location privacy valuation, including those of privacy attitudes, large questionnaires need to be administered, and for that online studies are a better suiting method than mobile crowdsourcing. To examine the impact of privacy attitudes, an online study was conducted. The participants were presented with an offer to share their location during a period of 1 month. The sharing would happen within a subsequent study, during which the location is logged nearly continuously at the highest available accuracy. The offer is based on a reverse auction, where the participants compete to be included in the deal by offering increasingly low prices [52]. The participants are led to believe that 25% of them would be chosen as per the offer, and given actual compensation for the logged data. They would receive the amount of the highest bid placed among the lowest 25% of the bids; this way none of the accepted participants would be paid less than the bid they had placed. Bidding too high would result in not being chosen; on the other hand, bids are expected not to be lower than each user's level of comfort. Thus, the responses can be considered as real statements of location privacy valuations.

An online survey was conducted to assess how much value users attribute to their location information. Also the relationship of this valuation and location privacy attitudes, including privacy perceptions and knowledge, was assessed.

6.2.1   Method

The online survey ($N = 106$) was conducted using the LimeSurvey platform. Participants were recruited using Prometei, which is a participants' database provided by TU Berlin. The survey was estimated to take roughly 20 min, and the participants were offered an Amazon voucher of 5€ as compensation.

6.2.1.1   Measures

The independent variables measured within this study that have been presented earlier in Chap. 3 were RISK, BENE, PCAL, NORM, and KNOW. The measures that were not presented previously are described here. These include *location privacy valuation*, *LBS usage frequency*, and *prior privacy violations*.

Location Privacy Valuation (LVAL)
To evaluate how much the participants value their location data, we used reverse auction, a methodology used earlier in privacy context by Danezis et al. [40] and Cvrvek et al. [39]. The participants were asked to state a price that they would require to be paid as compensation in order to participate in a subsequent field study within which their location would be collected automatically and continuously (e.g. every 5 min) for a period of 1 week at the highest accuracy available. A reverse

auction was used to make sure the price estimates reflect the users' preferences; the participants were told that those with the lowest 25% of the bids could be taken into the subsequent study. No opt-out option was provided in order to collect opinions also from the participants who were not, for whatever reason, interested in participating in the subsequent study. Instead, it was stated that providing a value in the bidding did not oblige the participant for anything; they would not automatically participate in another study, and their information would not be shared without their explicit consent.

Responses showed a large variance, with the minimum values at 0€, and maximum at 500,000€. It can be assumed that, despite the attempt at clarifying that giving a bid is not a consent to participate in the subsequent study, the very high values were stated in the absence of an option to opt out. Thus, the highest bids can be considered outliers, and were removed from the data prior to the analysis ($N = 8$). After removing the outliers, the statistics (in €) are as follows: $M = 70.50, SD = 138.35, Median = 27.50$. For statistical analysis, the variable is binned into seven categories.

LBS Usage Frequency (USE)

The frequency of using location-based services was measured by asking the participants how often they use various applications and functionalities. Ten different applications or functionalities were listed. The question to enquire about the usage was worded as follows:

Q. *How often do you use the following features or services?*

- Localization for sports activities
- Finding services based on your location
- Navigation
- Finding information about nearby places
- Sharing your location with others
- Location-based applications to locate children and seniors
- Marking visited places as a remembrance
- Location-based applications for finding lost property
- Location-based services to arrange meetings
- Location-based games

Each of these was scored on a fully labelled four-point Likert scale: "Never" (0), "Rarely" (1), "Occasionally" (2), and "Often" (3). A mean value was calculated to get a score for LBS usage frequency, $M = 0.89; SD = 0.47$.

Prior Privacy Violations (VIOL)

Privacy violations experienced in the past were measured by asking participants whether or not they had experienced certain types of violations. These violations were particular for mobile usage, for instance, the participants were asked if they had experienced location-based advertising. They were also asked if they knew somebody who has been a victim of a privacy breach. All items were measured on a dichotomous scale "yes" (1) or "no" (0), and a mean value of the responses was calculated to obtain a score for past privacy violations. The items were as follows:

1. *Inappropriate Data Collection.* First, we asked the participants if they have ever felt discomfort because of data collection that had happened as a consequence of their usage of mobile phone (or online activities). The question was posed this way to identify and record only those incidents of behavioural data collection that have made the participants feel uncomfortable.
2. *Location-Based Adverts.* The category of location-based adverts includes emails, offers, and other adverts based on visited locations. Additionally, to assess whether or not the received adverts could be considered privacy violations, the participants who expressed having received location-based adverts were asked how they felt about these adverts. This was measured on a fully labelled five-point answer scale as follows: "Very comfortable" (0), "Comfortable" (1), "Neutral" (2), "Uncomfortable" (3), "Very uncomfortable" (4) ($M = 2.92$, $SD = 0.74$).
3. *Stalking.* We asked whether the participant had ever been a victim of stalking because of their mobile phone use.
4. *Robbery.* We asked whether the participant had ever been a victim of robbery because of their mobile phone use.
5. *Identity Theft.* We asked if the participant had ever been a victim of identity theft or impersonation because of their mobile phone use.
6. *Knowing a Victim of a Location Privacy Violation.* Finally, we asked the participants if anyone whom they know have been in a situation where that person's location privacy has been compromised.

6.2.1.2 Sample

Altogether 109 individuals participated in the study. The age and gender distribution of the study is presented in Fig. 6.10; only binary genders are considered in this statistic. The distribution of educations in the study is detailed in Fig. 6.11, and occupations in Fig. 6.12. Of the participants of this online study, 8.3% worked, or have worked in the past, in a field related to information technology.

6.2.2 Results

In this section, the conducted statistical analyses are presented briefly; the findings are discussed in more detail in Sect. 6.2.3. First, the reported privacy violations are discussed (Table 6.5). Then, the relationships between measured privacy attitudes are assessed, including how perceived risks and benefits relate to risk–benefit assessment, and subsequently, how this relates to location privacy valuation (for all correlations, see Table 6.6). Non-parametric correlations were used when assessing relationships involving PCAL, LVAL, and KNOW, because these variables are not

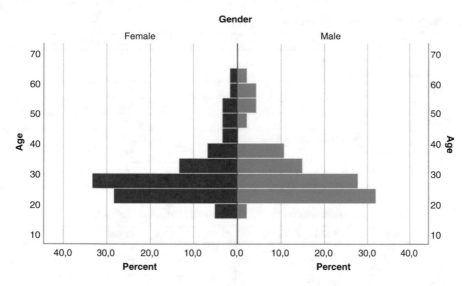

Fig. 6.10 Age and gender distribution of the Study V

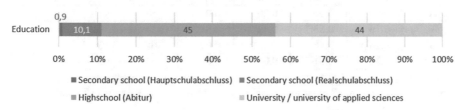

Fig. 6.11 Distribution of educations in Study V

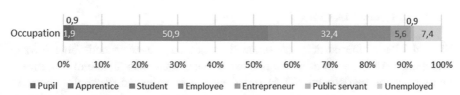

Fig. 6.12 Distribution of occupations in Study V

normally distributed ($Shapiro-Wilk = 99$, $p < 0.001$, for all variables). Next, the effect of some background factors, namely, that of past privacy violations as well as location privacy knowledge, on location privacy attitudes, is assessed. Finally, the influence of privacy attitudes on location disclosing behaviour measured as usage of location-based applications is evaluated.

Table 6.5 Reported privacy violations

Privacy violation	Percent of participants
Inappropriate data collection	77.1%
Location-based adverts	76.1%
Stalking	2.8%
Robbery	1.8%
Identity theft	9.2%
Knowing a victim of a location privacy violation	12.0%
Any privacy violation	92.7%

The most typically reported violations in this study were inappropriate data collection, and location-based advertising. Nearly all participants reported some privacy violations

Table 6.6 Correlation coefficients for the measures included in the online study

	RISK	BENE	PCAL	NORM	LVAL	USE	VIOL
RISK	1						
BENE	−0.16	1					
PCAL	0.43**	−0.48**	1				
NORM	−0.11	0.58**	−0.45**	1			
LVAL	0.25*	−0.06	0.20*	−0.04	1		
USE	−0.16	0.53**	−0.48**	0.34**	−0.14	1	
VIOL	0.26*	−0.16	0.35**	−0.09	0.39	0.53	1
KNOW	0.14	0.01	0.04	−0.06	0.12	−0.00	0.09

The correlations involving PCAL, LVAL, as well as KNOW were assessed using non-parametric Spearman's correlation, all other relationships were assessed using Pearson's correlation. $^*p <$ 0.05; $^{**}p < 0.01$

6.2.2.1 Privacy Violations

In total 93% of the participants stated that they have experienced some privacy violations in the past, all of whom had experienced them also first hand. Over three quarters of the participants (77%) stated having experienced discomfort because of data collected during mobile phone or internet usage. A similar proportion of participants (76%) reported that they had received behavioural adverts based on their location. All other types of violations were reported less (cf. Table 6.5).

The participants who stated having received location-based adverts considered them mostly discomforting; 71.1% reported having felt either "uncomfortable" or "very uncomfortable" after receiving such adverts.

6.2.2.2 Privacy Attitudes

The relationships between privacy attitudes were evaluated. A positive medium correlation was found between risk perception and privacy calculus, and a negative one between benefit perception and privacy calculus. To investigate the relationship between these measures further, an ordinal regression was conducted. The results from the regression analysis indicate that an increase in the RISK score is associated with an increase in the odds of estimating that risks outweigh the benefits in the usage of LBS with an odds ratio of 0.62 (95% CI : [0.32, 0.91]); whereas, an increase in the BENE score is associated with an increase in the odds of estimating that risks outweigh the benefits with an odds ratio of -0.91 (95% CI : $[-1.26, -0.56]$). This result gives a good indication that privacy calculus measured as PCAL in this study does correspond to the expected relationship of perceived risks and benefits.

In order to assess whether location privacy valuation can be considered a quantification of privacy calculus, the relationship of PCAL and LVAL was analysed using Spearman's non-parametric correlation. The results indicate a small positive correlation between the two measures, suggesting that location privacy valuation does to some extent relate to privacy calculus in the context of privacy. To further evaluate this relationship, a linear regression was conducted. The results show that the model was significant ($F(1, 107) = 4.29, p = 0.041$), and privacy calculus significantly contributes to predicting privacy valuation ($\beta = 0.32, p = 0.041$), however, the model explains only 4% of the variance in privacy valuation ($R^2 = 0.04$).

6.2.2.3   Predicting Privacy Attitudes

The influence of prior privacy violations on privacy attitudes was evaluated. A positive correlation was identified between VIOL and RISK, as well as between VIOL and PCAL, suggesting that prior privacy violations might influence the way users perceive risks, and also the tendency to consider that risks outweigh the benefits in the usage of LBS—possibly mediated by risk perception. Mediation is a causal chain where one variable influences, or "mediates" the relationship between two other variables; mediation can be determined using a series of three simple regression models [22]. Thus, this mediation effect was assessed using regression analysis under the assumption that PCAL can be treated as a continuous variable. The results indicate that the relationship between VIOL and PCAL is mediated by RISK: as illustrated in Fig. 6.13, the standardized regression coefficient between VIOL and RISK was statistically significant, and likewise, the standardized regression coefficient between RISK and PCAL was statistically significant. The contribution of RISK on PCAL was diminished when controlled for RISK, which is consistent with a partial mediation assumption. A significant indirect effect of VIOL on PCAL through RISK was identified using a bootstrapping approach with 5000 samples; $b = 0.86$ (95% CI: [0.14,1.98]). The result suggests that the prior privacy violations do have an influence on to what extent risks are considered to outweigh the benefits in the usage of LBS, mediated by perceived risks.

The influence of education on privacy attitudes was evaluated using a non-parametric correlation. A small positive correlation was identified between level of education and perceived risks ($r_s = 0.28, p < 0.05$), and between level

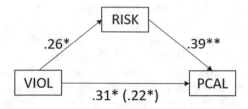

Fig. 6.13 Standardized regression coefficients for the relationship between prior privacy violations (VIOL) and privacy calculus (PCAL) as mediated by perceived risks (RISK). In parenthesis
is the standardized regression coefficient between VIOL and PCAL, controlling for RISK. $^*p <$
$0.05, ^{**}p < 0.01$

of education and privacy calculus ($r_s = 0.23, p < 0.05$). Also differences
in privacy attitudes between genders were identified: women perceived more
risks than men ($t(108) = 2.23, p < 0.05$), perceive less benefits than men
($t(108) = -2.24, p < 0.05$), and find more often than men that risks outweigh the
benefits in the usage of LBSs ($t(108) = 2.28, p < 0.05$). No differences based on
age of the participants were identified.

Finally, there was no relationship found between KNOW and any other variable
measured in this study.

6.2.2.4 Predicting Intention to Disclose

The influence of attitudinal factors on self-reported usage frequency was evaluated.
No correlation was identified between LVAL and USE, implying that location
privacy valuation and usage frequency of LBS are not connected. A positive
correlation was found between USE and BENE as well as with NORM, and a
negative correlation between USE and PCAL. To further assess the relationship
between these factors, a linear regression analysis was conducted. The results
suggest that PCAL explains 22% of the variance in USE; however, when BENE
is also added to the model, the two variables together explain 31% ($F(2, 106) =$
$25.26, p < 0.001; R^2 = 0.31$). Adding NORM to the model did not improve the
prediction ($R^2 = 0.30$), NORM also did not contribute significantly to the model
($p = 0.617$). LBS usage frequency (USE) can be expressed as a function of privacy
calculus (PCAL) and perceived benefits (BENE) as the following equation:

$$USE = 7.45 - 0.11 \times PCAL + 0.15 \times BENE. \qquad (6.2)$$

6.2.3 Intermediate Discussion

This study was conducted to overcome the drawback of the crowdsourcing study
that only limited sized questionnaires could be administered using the mobile

crowdsourcing platform. Thus, this study was conducted as an online questionnaire; various beliefs and attitudes around location privacy were included in the questionnaire, as well as a self-reported measure for usage frequency of LBS. Additionally, two background factors were included in the study: location privacy knowledge and prior privacy violations.

6.2.3.1   Privacy Violations and Privacy Attitudes

First, the relationship between privacy calculus and perceived risks as well as perceived benefits was analysed—the results confirmed the expectations that privacy calculus is influenced both by the risks and by the benefits that a user perceives in using LBS. Additionally, prior privacy violations were found to influence this calculus, mediated by perceived risks. The mediation means that experiencing privacy violations increases the likelihood of increased risk perception, which leads to a perception that risks outweigh benefits in LBS. The privacy violations measured in this study included, among other violations, having experienced inappropriate data collection, or location-based adverts. Having experienced privacy violations in the past does not seem to correlate with the amount of benefits perceived in LBS.

Most participants reported having experienced some privacy violations, of which the most typically reported ones were inappropriate data collection, and location-based adverts, each of which were mentioned by three quarters of the participants. It should be noted, however, that this number comprises only the participants who have paid attention to the fact that they have received adverts that are based on their location information—in reality the number is possibly higher. However, this result raises a question about the ethicality of behavioural advertising: if users who do notice that they are receiving adverts based on locations they have visited feel mostly uncomfortable about the adverts, can the financial benefits of the advertiser—or the benefits received by the user through well-targeted service—justify the caused discomfort?

6.2.3.2   Location Privacy Knowledge

No relationship was identified between location privacy knowledge and any other variable measured in this study. This result is interesting, as it could be expected that increased knowledge in how location information is handled would have an influence on perceived risks. One explanation to why no relationship is found could be that the knowledge influences different users in different ways, such as:

1. A user might, because of increased knowledge, understand the risks and thus thinks that they can be dealt with, and possibly also consider that the risks are not overbearing the benefits.

2. A user might, because of increased knowledge, know about the risks and consider them perilous—and possibly, consider the benefits not to be strong enough to outweigh the risks.

Another possibility is that the measure for location privacy knowledge as measured in this work is not adequate for assessing the spectrum of general knowledge in how consumers' location information is collected and used.

6.2.3.3   Location Privacy Valuation

In this study, a positive relationship was identified between privacy calculus and location privacy valuation, suggesting that users who consider risks to outweigh the benefits in the usage of LBS also attribute more monetary value on their location information. However, the variance in location privacy valuation (LVAL) explained by privacy calculus (PCAL) was too small to conclude that LVAL as measured in this study would be an adequate metric for quantified privacy calculus.

No correlation was identified between location privacy valuation and self-reported usage frequency. This could be also due to the nature of measuring valuation in this study, which considered value of the potential location data collected in the future. As the data collection has not happened yet, the participants might have difficulties assessing the extent to which sensitive information might be included in the location data, and therefore a proper assessment of value for the information becomes challenging.

6.2.3.4   Predicting LBS Usage Frequency

For estimating privacy behaviour as usage frequency of location-based applications, privacy calculus and perceived benefit were found to provide the best prediction power; nearly one-third of the variance in usage of such applications could be explained by the two variables. Higher amount of benefits seen in the usage of LBS increases the likelihood that such applications are used, as well as does the prospect that benefits outweigh risks in their usage. Social norm—or how much influence peers' opinions and expectations have—has a positive relationship with using location-based applications, however, its effect in the model predicting the behaviour remained non-significant.

Removing Outliers
Cases with high values of location privacy valuation were removed from the data. This might have led to elimination of the effect of high privacy concern, as the high values might represent the users who are not willing to give their location data away, or are willing to do so only at a very high compensation. However, the test procedure required the participant to state a value at which they would be willing to participate in a month-long data gathering, and a high value is merely indicating unwillingness to participate. This might be due to privacy concern, but

high values could have been stated also by users who were unwilling to participate for other reasons, such as being busy. This assumption is affirmed by assessing the distribution of the perceived privacy risks in the high valuation group in comparison to the rest of the dataset. No differences were found using a Mann–Whitney U-test, suggesting that the highest location privacy values were not solely due to privacy concern ($U = 340.50$, $p = 0.243$). Therefore, it can be assumed that the participants attributing extremely high values for location privacy can be justifiably disregarded from the dataset.

6.3   Chapter Summary

The aim in this chapter was to evaluate whether or not privacy calculus in the context of location could be quantified through monetary quantification, and whether this metric could be used to estimate location privacy behaviour.

The findings from the two studies presented in this chapter offer to some extent contradicting results with respect to the adequacy of location privacy valuation as a method to estimate location disclosure. In the Study IV presented in the first part of this chapter, Sect. 6.1, location valuation was found to function rather well for estimating location sharing in the crowdsourcing scenario—in particular when considered in combination with trust. However, in the study V, location valuation did not seem to predict privacy behaviour. The contrasts in these findings could be explained by the differences in how the location privacy valuation was assessed, but also in the type of behaviour that was evaluated.

Chapter 7
Privacy Protection Behaviour

7.1 Study VI: Which Factors Influence Privacy Protection Behaviour?

This chapter presents a study providing users a possibility to obfuscate their location information in order to protect themselves from privacy breaches using a simple and illustrative settings application. Adoption of different privacy protection mechanisms has been previously examined in various contexts. Lindqvist et al. found that users might adopt various techniques to improve their locational privacy, including not sharing their location with an application, and not connecting data between social media sites [100]. In the context of protection of location privacy, Wei et al. suggest a system providing more flexibility for the user to control their location disclosures for mobile social networks [173]. Brush et al. investigated users' preferences for location obfuscation [32]. The results from their study suggest that users can comprehend the impact of obfuscation methods on their location privacy, and are motivated to choose obfuscation methods according to their privacy preferences.

The influence of privacy nudges on the usage of location protection behaviour is also assessed. The impact of privacy nudges on users preferred privacy settings has been assessed before in empirical studies. Users of LBS have been found to be unaware of the data that is collected through the applications that they use, and informing them using privacy nudges prompts to re-evaluate and restrict permission given to applications [3, 10].

This section presents an empirical study ($N = 48$) assessing users' location privacy protection behaviour in a real-life setting. The participants of the study are offered a software for controlling their location privacy using different methods. The field study is accompanied by a large questionnaire to evaluate the relationship of participants' privacy attitudes and usage of the location privacy protection mechanisms. Ultimately, the users' privacy behaviour should be directed towards

© Springer Nature Switzerland AG 2020
M. E. Poikela, *Perceived Privacy in Location-Based Mobile System*,
T-Labs Series in Telecommunication Services,
https://doi.org/10.1007/978-3-030-34171-8_7

more informed, safer choices. One of the goals of this part of the work is to achieve this through system design. Therefore, a solution was implemented for increasing users' awareness of their location sharing on a day-to-day basis. The users were informed about their location sharing through notifications called *privacy nudges*.

7.1.1   Method

A 15-day field study was conducted to evaluate the relationship between privacy attitudes and actual privacy protection behaviour. The study was conducted between November 2017 and March 2018 with German participants. Participants for the study were recruited using a participants database Prometei provided by TU Berlin, and through online adverts. Additionally, participant recruitment was attempted through mailing lists for computer scientists in order to find participants with devices compatible with the test application. This attempt was, however, not successful. Thus, all the participants used test devices provided for them. Because of limitations posed by the number of test devices, the study was repeated multiple times in order to acquire a sufficient number of participants. The participants' efforts were estimated to be in total a maximum of 3 h during the 2 weeks, for which they were compensated with an Amazon voucher of 30€. The study received ethical clearance from an ethical committee of TU Berlin.

The study had three phases: an introductory session, a field study, and a final meeting (cf. Fig. 7.1). These phases are presented here in detail.

Introductory Session
Within the introductory session (30 min) the participants were given an explanation to the study, its purposes, and the details regarding data collection. Next, they filled out a demographic questionnaire together with a consent form. In addition, questions regarding behavioural intention were included in this questionnaire. The participants then signed a lending agreement, and received a test device with the Protect Location application pre-installed. Also WhatsApp, Facebook, and an application for the local public transport system were pre-installed on the test devices. The participants were given an introduction to the functionalities of the software, called Protect Location, after which the test device was set up according to the participants' preferences. Their SIM card(s) and a possible SD card were transferred from their own device to the test device. Any applications they needed were installed either within this session, or the participants were encouraged to do so at their own peace afterwards. The participants were also guided to sign in to the applications that are important for them, and in many cases, a WhatsApp chat history was archived on their personal device, and uploaded on the test device. The participants were requested to use the test device as their main phone during the field study. The participants were told that they can quit the study at any point, in which case they would receive a partial compensation. Finally, they were reminded

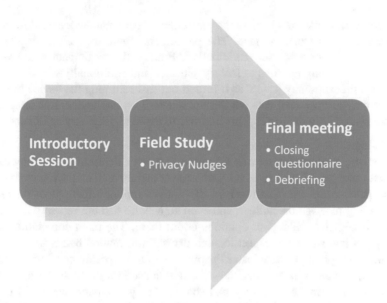

Fig. 7.1 The study consisted of three parts: (1) an introductory session, during which a demographic questionnaire was filled out, an introduction to the experimental application was given, and a test device was set up; (2) a field study, during which a random 60% of the participants received privacy nudges; and (3) a final meeting, which included a closing questionnaire including a scale for privacy attitudes and beliefs, as well as a debriefing

that the research team would be happy to answer any questions they might have during the study.

Field Study
During the field study the participants had no specific tasks. On the contrary, they were instructed to behave as normally as possible in order to make as realistic setting as possible to assess the use of the Protect Location application—to what extent it is used in a real-life scenario, and how. During the field study, 60% of the participants received privacy nudges that were specific to their home location, work address, the places they had visited, or the applications that had used location on their device, while the remaining 40% were a control group.

Final Meeting
Within the final meeting (30 min) the participants filled out a final questionnaire including attitudinal items and other measured variables (*perceived benefits, perceived risks, risk–benefit assessment, social norm, past privacy violations, privacy concern,* as well as *trust*). Next, a debriefing was given regarding data collection and anonymity, including a mention that the location data that was collected within the study was only collected for sending the privacy nudges—also the purpose of the nudges for this research was explained. The participants were asked if these issues were adequately clarified to make them feel at ease, and if they had any further questions. Then, the participants' personal SIM card(s), and if applicable,

SD card, were transferred back to the participant's personal devices and made sure that everything is working again. The participants were then given the Amazon voucher as compensation for participation. Finally, the participant could witness the test device being returned to factory settings, and additionally wiped. This was not only a precautionary action, as it turned out that returning the device to factory settings only hid personal data but did not delete it on the device; this information was not shared with the participants in order to avoid unnecessary anxiousness.

Debriefing

In this study, each participant was individually given a debriefing regarding the use of privacy nudges. In most cases the participants commented on the topic of nudges: several participants who had received the nudges reported feelings of discomfort when having received the nudges, and even some who had not received the nudges expressed negative feelings or opinions about them. The main complaint that the participants had was with respect to such private information being revealed. Care was taken to explain the participants the purpose of the experimental design and data collection, as well as the fact that the information used for the nudges was based on the location information that they had shared. The participants were told that the location data will be permanently deleted after the study, and that if they had any questions or concerns even after the study, they could contact the research team at any point. Comfort or discomfort with respect to the received privacy nudges was not explicitly measured within this study.

7.1.1.1 Apparatus

For the study, a location settings application called "Protect Location" was created. The application gives the user several options for controlling location privacy by letting the user decide the accuracy at which each application on the device can access the physical location of the device. The Protect Location application works between the system location and the application layer, and provides the location information to other applications on the device. The user has four options for managing their location settings:

1. *Sharing the exact location* (with the highest accuracy provided by the network and GPS)
2. *Sharing custom location.* Using this option, the user can drop a pin on the provided map or search for a custom, or "faked", location. Sharing custom location, as well as sharing obfuscated location with a chosen accuracy, are illustrated in Fig. 7.2.
3. *Sharing obfuscated location* with a chosen accuracy. The user can select a radius (0 m to 5000 km), which is also displayed on the map. The location that in this case is shared with the applications on the device is selected randomly within this area.
4. *Denying location sharing*, which switches the location sharing off.

(a)   (b)

Fig. 7.2 Two screenshots of the Protect Location application illustrating the two available obfuscation methods: sharing a custom location (**a**), and sharing a chosen accuracy, which the user can select using a stepless slider between [0 m and 5000 km] (**b**). Alternatively, the user could share an exact location, or deny location sharing

Each of the location sharing settings could be chosen for any application on the device separately, or as a default setting for all applications. In the case that a default setting has been chosen, exceptions can be set for any application by selecting any setting for that application separately.

The Protect Location application needs to have root access to the Android operating system code to be able to modify the location information provided by the system before sharing it with the other applications on the device. Therefore, the devices that the participants were to use within the field study had to be *rooted*. Rooting one's device is not typical for a normal user, as doing so requires advanced know-how, and likely also affects the warranty of the device. Thus, it could not be assumed that naïve users with suitable devices could be recruited. Users with rooted devices were sought; however, with the assumption that such users would deviate from others in knowledge, and possibly also in attitudes and behaviours. In order to conduct the study with naïve users, altogether 18 Android devices of the model Thor from the brand Vernee, which were suitable for the study, were purchased.

7.1.1.2   Measures

Various variables were measured during the study for assessing the background factors, privacy attitudes, as well as behaviours. The independent variables included in this study that have been presented in Chap. 3 were RISK, BENE, PCAL, NORM, and WILL. The measure for *Prior Privacy Violations* (VIOL) has been detailed in Sect. 6.2. *Location Privacy Valuation* (LVAL), which is measured similarly to that in Study V (Sect. 6.2), as well as *Distrust* (DTRT), which is measured similarly to the measure for trust in Study IV (cf. Sect. 6.1), are detailed in this section.

To measure behaviour, three types of disclosing behaviour (*intention to use LBS*, *installing applications*, and *sharing location*) as well as three types of protection behaviour (*uninstalling applications*, *custom location*, and *switching location off*) were measured. The first 3 days of the field study were assigned as an acclimatization period, during which the participants get accustomed with the new device and experiment with the various location privacy protecting functionalities out of curiosity. To omit the influence of the acclimatization period, the protection behaviour is considered only after the first 3 days.

Privacy nudges were used to increase participants' awareness of their location information disclosures. Such awareness, or the impact of privacy nudges on it, was not explicitly measured; the nudges are described in this section.

Location Privacy Valuation (LPV)
Location privacy valuation was measured in a similar fashion as in the online study presented in Sect. 6.2, with the difference that within the field study the participants had to assess the value that they would attribute to location information after the information was already collected. They were asked how much they would need to be paid for their location data from the "past 2 weeks" to provide the information to a third party, which was presented as a startup named "Standort Berlin". As in the online survey, a method of reverse auction was used in order to find the lowest value that the participants would be comfortable in accepting for the data disclosure; the participants whose bids were within the lowest 25% would be considered.

Distrust (DTRT)
To measure trust, the participants were asked to what extent they trust or distrust the startup "Standort Berlin", which was presented as a receiving party for the collected location data. Because no information was provided regarding the company and therefore the participants did not have any prior attitudes towards the company, this trust score is speculated to reflect the participants' personalities. Trust towards the startup was measured on a four-item scale using a fully labelled seven-point Likert scale (0–6) similar to one used in Study IV (cf. Sect. 6.1). As a comparison, trust towards the Quality and Usability Lab (QU Lab) of TU Berlin was measured using the same scale. Since the participants, participating in a study, had already agreed on sharing some personal information with the QU Lab, the participants were expected to have higher trust towards the QU Lab than the startup. A paired samples t-test was conducted to compare the level of trust towards the QU Lab and towards the startup. Significantly less trust was expressed towards the startup ($M = 4.23; SD = 1.27$) than the QU Lab ($M = 5.66; SD = 0.92$); $t = -6.91, p < 0.001$. This suggests that the QU Lab was indeed considered as rather trustworthy, whereas more feelings of distrust are present towards the unknown data collector, the Standort Berlin startup.

Only the negatively coded items were included in the analysis, and therefore the measure gives an assessment of *distrust*, which is the extent to which a participant finds an entity to be untrustworthy and potentially even causing harm.

Disclosing Behaviour

Altogether three types of disclosing behaviour were measured: *intention to use LBS*, *installing applications*, and *sharing location*.

Intention to Use LBS The participants were asked how likely they thought it was that they would use various location-based functions or services during the following 2 weeks. Based on the findings of the interview study presented in Sect. 4.1, ten types of functionalities or applications were considered, such as navigation, location-based games, sharing one's location with others, and finding services based on location. The question to enquire the participants about their intention to use LBS was worded as follows:

Q. *How likely are you to use the following features or services over the next 2 weeks?*

- Localization for sports activities
- Finding services based on your location
- Navigation
- Finding information about nearby places
- Sharing your location with others
- Location-based applications to locate children and seniors
- Marking visited places as a remembrance
- Location-based applications for finding lost property
- Location-based services to arrange meetings
- Location-based games

Each of the items was rated for probability of usage on a fully labelled Likert scale: "Very unlikely" (0), "Unlikely" (1), "Likely" (2), and "Very likely" (3); $M = 2.22, SD = 0.51$.

Installing Applications The number of applications that the users install on their devices per day during the 2 week period was recorded. This number includes the times that a specific application is uninstalled, and installed again ($M = 2.93, SD = 0.72$).

Sharing Location The number of times that a participant switches location sharing on per day by using the setting *Sharing the exact location* in the Protect Location app is measured ($M = 0.39, SD = 0.07$).

Protection Behaviour

Four types of protection behaviours were measured during the study: *uninstalling applications*, *custom location*, *obfuscated location*, and *location off*. However, only three of these, namely *uninstalling applications*, *custom location*, and *location off* could be used for analysis because of scarcity of data.

Uninstalling Applications The participants could uninstall any applications they wished on the smartphone (except the Protect Location app, which the participants were requested to keep). The number of uninstalled applications per day was recorded as a privacy protection measure ($M = 0.20$, $SD = 0.10$).

Custom Location Custom location refers to number of times per day that the participant has chosen the setting *sharing custom location* in the Protect Location app ($M = 0.13$, $SD = 0.04$).

Obfuscated Location The number of times per day that a participant selects the option *sharing obfuscated location with a chosen accuracy* in the Protect Location app is measured ($M = 0.01$, $SD = 0.01$). The setting was used very little, and as a consequence, this type of protection behaviour is not considered in the further analyses.

Location Off The number of times per day that a participant switched the location setting off was measured ($M = 0.01$, $SD = 0.01$).

Awareness Through Privacy Nudges
After the 5th day of the field study, a randomly selected 60% of the participants started receiving privacy nudges, which were created based on the location information that the participants had disclosed; the remaining 40% were a control group. Based on the information available because of the participants' location sharing settings, their home addresses and work addresses were automatically calculated; the nudges informed the participants that this information was available because of the location information that they shared. Determining the location of home and work was simplified, based on recording the user's physical location during the night for determining the home address, and during office hours for determining the work address. This method has limitations, including not considering users working night shifts, or users who move around during their work, such as drivers, or security personnel. In addition to the information regarding home and work locations, the number of applications that have used the participant's location information was recorded and used for a privacy nudge.

The location information used to generate the privacy nudges was based on the information that the participants shared on their device. For example, had the participant shared a custom location of Easter Islands, a privacy nudge sent to that particular participant could state that their work address is on Easter Islands. As examples of the used privacy nudges, the following messages were sent to some participants:

- "Facebook and 3 other applications have requested your location today: Facebook (131 times), Maps (10 times), Chrome (6 times), Messenger (5 times)."
- "You visited the restaurant Soup Cult today."

The influence of privacy nudges is measured implicitly by comparing how the experimental group, which received nudges, differs from the control group in attitudes and behaviour.

7.1.1.3   Sample

The study was advertised for users with rooted Android devices who could install Protect Location directly on their devices, and additionally to any participant who would be willing to use a test device for the duration of the study as a primary device. No participants with rooted devices were found for the study; all the participants received test devices.

Altogether 54 participants took part in the study; six did not finish the study either because they were unhappy about the test device or its compatibility with services important to the user, they could not use mobile data on the device for technical reasons, or for other personal reasons. The following analysis is based on data from the remaining 48 participants.

The participants were mainly young adults ($M = 32.25\ yrs$, $SD = 9.77$), and a majority of the participants were female (65%). The gender and age distribution of the study is presented in Fig. 7.3; only participants with a binary gender have been considered in this statistic. The participants were relatively highly educated, with 86% having at least a high school degree, and 40% having some university degree; the education distribution within the study participants is detailed in Fig. 7.4. Half of the participants were students (51%); the occupation distribution is detailed in Fig. 7.5. In total 18.8% of the participants of this study stated that they work, or have worked in the past, in a field related to information technology.

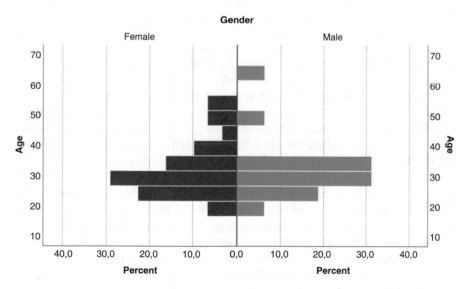

Fig. 7.3 Age and gender distribution of the Study VI

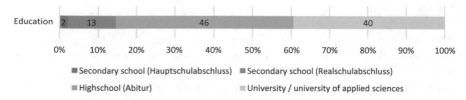

Fig. 7.4 Distribution of educations in Study VI

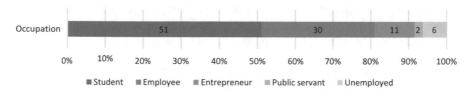

Fig. 7.5 Distribution of occupations in Study VI

7.1.2 Results

In this section, the statistical results are assessed. The results are later discussed in detail in Sect. 7.1.3. First, the relationships between background factors and privacy attitudes that have not been addressed in previous chapters are evaluated. Then, the location disclosing and location protection behaviours are assessed, as well as how these are influenced by the privacy attitudes. Then, the influence of these attitudes on the intention to use LBS and stated usage frequency of LBS is assessed, as well as how the intention relates to location disclosing behaviour. Finally, the impact of privacy nudges is discussed. All correlation coefficients for the independent variables measured in the study—location privacy attitudes, as well as background factors—are reported in Table 7.1. Several variables were not normally distributed, and therefore non-parametric correlations were conducted when assessing relationships with these variables; these include PCAL ($Shapiro–Wilk = 0.899, p = 0.001$), VIOL ($Shapiro–Wilk = 0.87, p < 0.001$), KNOW ($Shapiro–Wilk = 0.803, p < 0.001$), and LVAL ($Shapiro–Wilk = 0.014, p = 0.014$). Also none of the behavioural measures are normally distributed; non-parametric correlations were computed. There were no issues identified with the normal distribution of the residuals in the regressions, homoscedasticity, or with multicollinearity [36].

For evaluating the relationships between background factors and privacy attitudes, the relationships of privacy attitudes and willingness to disclose information (WILL) as well as that of privacy attitudes and distrust (DTRT) have not been discussed in earlier chapters, and are addressed here. WILL has a positive correlation with BENE and NORM, and a negative correlation with PCAL. To assess these relationships further, simple linear regressions were conducted. According to the results, roughly quarter of the variance in social norm (NORM) can be explained

Table 7.1 Correlation coefficients for all independent variables measured in the field study, which include location privacy attitudes (RISK), (BENE), (PCAL), (NORM), and (LVAL), as well as background factors (WILL), (DTRT), (VIOL), and (KNOW)

	RISK	BENE	PCAL	NORM	WILL	DTRT	VIOL	KNOW	LVAL
RISK	1								
BENE		1							
PCAL	0.37*	−0.57**	1						
NORM		0.59**	−0.50**	1					
WILL		0.59**	−0.46**	0.50**	1				
DTRT	0.48**	−0.46**	0.35*		−0.43**	1			
VIOL	0.33*		0.31*				1		
KNOW			0.30*			−0.43**		1	
LVAL	0.43**		0.35*			0.38**	0.30*		1

$*p < 0.05$; $**p < 0.01$. For clarity, only significant correlations are reported

Table 7.2 Linear regression analysis predicting social norm (NORM) from willingness to disclose (WILL)

Model	Adjusted R^2	F	Predictor	B	SE B	β	t
1	0.26	17.63**	Constant	1.34	0.33		4.03**
			WILL	0.46	0.11	0.53	4.20**

$**p < 0.01$

by willingness to disclose personal information (WILL), see Table 7.2 for the regression results.

Following the taxonomy as postulated in Sect. 2.2, the relationship between WILL, BENE, and PCAL could be a mediation, which would imply a causal relationship between willingness to disclose personal information and perceived benefits, as well between perceived benefits and privacy calculus. This was evaluated with three simple regression analyses; the regression paths and coefficients are presented in Fig. 7.6. When controlling for perceived benefits, WILL was no longer a significant predictor of PCAL, which is in line with a full mediation hypothesis. The indirect effect was tested using bootstrapping with 5000 samples; the results indicated that the indirect coefficient was statistically significant, $B = -0.18$ (95% $CI : [-0.34, -0.07]$). Thus, willingness to disclose personal information was associated with stronger perception that benefits outweigh risks when mediated by perceived benefits.

DTRT has a negative correlation with BENE, and a positive correlation with RISK, PCAL, and LVAL. It could be postulated that the path of relationships is such that distrust has an influence on risk perception as well as benefit perception, both of which mediate the relationship with privacy calculus, which in turn affects location privacy valuation. This was first evaluated by assessing the simple mediation models: DTRT predicting PCAL mediated by BENE, and separately, DTRT predicting PCAL mediated by RISK. Based on the latter analysis, no mediating relationship was identified. Distrust does seem to influence risk

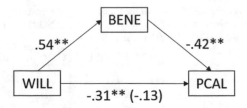

Fig. 7.6 Standardized regression coefficients for the relationship between willingness to disclose information (WILL) and privacy calculus (PCAL), mediated by perceived benefits (BENE). The standardized regression coefficient between WILL and PCAL, controlling for BENE, is in parenthesis; the path is no longer significant, which suggests a full mediation. ** $p < 0.001$

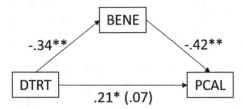

Fig. 7.7 Standardized regression coefficients for the relationship between perceived distrust towards data recipients (DTRT) and privacy calculus (PCAL), mediated by perceived benefits (BENE). The standardized regression coefficient between DTRT and PCAL, controlling for BENE, is in parenthesis; the path is no longer significant, which suggests a full mediation. * $p < 0.05$; ** $p < 0.001$

perception, $(R^2, F(1, 46) = 13.07, p = 0.001; \beta = 0.36, p = 0.001)$; however, no indirect effect on privacy calculus through risk perception could be identified. There was a relationship between perceived distrust and privacy calculus suggesting full mediation by perceived benefits. The regression coefficients for the relationships between the three variables are presented in Fig. 7.7. A bootstrapping procedure with 5000 samples was conducted for testing for the indirect effect from DTRT to PCAL through BENE $\beta = 0.14$ (95% $CI : [0.03, 0.29]$). This suggests that distrusting feelings were associated with a stronger feeling that risks outweigh the benefits when mediated by benefits; users feeling distrust towards recipients of location information are less likely to find benefits in LBS, and subsequently, perceiving less benefits increases the odds of feeling that risks outweigh benefits in the usage of LBS. Privacy calculus predicts location privacy valuation significantly, explaining 16% of the variance $(R^2 = 0.16, F(1, 46) = 9.05, p = 0.004; \beta = 0.69, p = 0.004)$. Thus, these results seem to indicate a causal pathway from distrust to privacy calculus through perceived benefits, and from privacy calculus to location privacy valuation.

Additionally, in contrast to the results presented in Sect. 6.2, KNOW has a positive correlation with PCAL. A simple regression analysis was conducted to evaluate this relationship further. The model did not turn out to adequately predict PCAL $(Pearson \quad \chi^2 = 5.04, p = 0.83)$; thus, as previously noted, the adequacy

Table 7.3 Correlation coefficients for the behavioural measures

	Intention to use LBS	Installing applications	Sharing location	Uninstalling applications	Custom location	Location off
Intention to use LBS	1	0.32*				
RISK						
BENE	0.30*				−0.32*	−0.40**
PCAL	−0.29*				0.29*	0.32*
NORM		0.46**	0.36*	0.42**		−0.33*
WILL	0.44**		0.37*			−0.36*
DTRT	−0.34*					0.36*
VIOL						
KNOW	−0.30*					0.38**
LVAL						

$^*p < 0.05$; $^{**}p < 0.01$. For clarity, only significant correlations are reported

of the measure for location privacy knowledge for predicting privacy attitudes could not be verified.

All correlational relationships between behavioural measures, and predictive variables including background factors as well as attitudinal factors, are listed in Table 7.3.

7.1.2.1   Disclosing Behaviour

Disclosing behaviour was analysed in terms of intention to use LBS, installing applications, and sharing location. These are described here in relation to privacy attitudes and knowledge.

Intention to Use LBS First, the relationships between attitudinal factors and intention to use LBS were evaluated based on Spearman's rank-order correlation results, detailed in Table 7.3. A simple regression analysis revealed that WILL predicts *intention to use LBS*, explaining 24% of the variance. BENE, PCAL, and DTRT work as predictors to a lesser extent, explaining, respectively, 11%, 7%, and 12% of the variance. KNOW was not found to predict *intention to use LBS*. The simple regression results are reported in Table 7.4. A multiple linear regression suggests that WILL works as a predictor for *intention to use LBS*, whereas the other variables did not contribute to the model significantly.

Installing Applications *Installing applications* correlates positively with *intention to use LBS*, and with NORM (cf. Table 7.3). No problems with collinearity were identified (the acceptable values according to Hair et al. are $Tolerance < 0.02$, $VIF < 5$ [70]). According to simple linear regression analyses, *installing applications* can be predicted from *intention to use LBS*, and from NORM. However,

Table 7.4 Summary of linear regression analysis predicting intention to use LBS

Model	Adjusted R^2	F	Predictor	B	SE B	β	t
1	0.11	6.94*	Constant	2.22	0.07		32.15**
			BENE	0.18	0.07	0.36	2.64*
2	0.07	4.85*	Constant	2.18	0.07		30.03**
			PCAL	−0.18	0.08	−0.31	−2.20*
3	0.24	14.21**	Constant	2.22	0.07		34.28**
			WILL	0.25	0.07	0.49	3.77**
4	0.12	6.92*	Constant	2.22	0.07		31.24**
			TRST	0.19	0.07	−0.37	−2.63*
5	−0.04	0.10	Constant	2.22	0.10		22.32**
			KNOW	0.03	0.10	0.06	0.31

*$p < 0.05$; **$p < 0.01$

Table 7.5 Summary of linear regression analysis predicting installing applications

Model	Adjusted R^2	F	Predictor	B	SE B	β	t
1	0.07	4.55*	Constant	16.82	1.67		10.10**
			Intention	3.59	1.68	0.30	2.13*
2	0.08	5.19*	Constant	16.82	1.67		10.07**
			NORM	3.85	1.69	0.32	2.28*
3	0.11	3.98*	Constant	16.82	1.64		10.24**
			Intention	2.77	1.72	0.23	1.61
			NORM	3.12	1.72	0.26	1.82

*$p < 0.05$; **$p < 0.01$

Table 7.6 Summary of linear regression analysis predicting sharing location

Model	Adjusted R^2	F	Predictor	B	SE B	β	t
1	0.10	6.07*	Constant	0.61	0.07		8.51**
			NORM	0.18	0.07	0.35	2.46*
2	0.11	6.47*	Constant	0.61	0.07		8.54**
			WILL	0.19	0.07	0.36	2.54*

*$p < 0.05$; **$p < 0.01$

these variables explain only, respectively, 7% and 8% of the variance in *installing applications*. Running a multiple regression analysis, the model turned out to be significant; however, neither of the predictors did (cf. Table 7.5). It can be concluded that NORM is the best predictor for *installing applications*, predicting the behaviour slightly better than *intention to use LBS*.

Sharing Location Location sharing can be predicted from perceived norm, or from willingness to disclose location. These models explain, respectively, 10% and 11% of the variance in sharing location (cf. Table 7.6).

Table 7.7 Linear regression analysis predicting uninstalling applications

Model	Adjusted R^2	F	Predictor	B	SE B	β	t
1	0.06	4.23*	Constant	0.97	0.11		8.71***
			NORM	0.23	0.11	0.29	2.06*

*$p < 0.05$; **$p < 0.01$

Table 7.8 Summary of linear regression analysis predicting usage of location obfuscation by custom location

Model	Adjusted R^2	F	N	Predictor	B	SE B	β	t
1	0.003	1.14	48	Constant	0.19	0.07		2.63*
				BENE	−0.08	0.08	−0.16	−1.07
2	0.09	5.48*	48	Constant	0.24	0.07		3.25**
				PCAL	0.19	0.08	0.34	2.34*
3	0.19	8.48**	36	Constant	0.25	0.09		2.81**
				PCAL	0.318	0.11	0.46	2.91**

*$p < 0.05$; **$p < 0.01$

Protection Behaviour

Protection behaviour is evaluated in terms of uninstalling applications, using the obfuscation method of custom location, and switching location off; the results from these analyses are presented here.

Uninstalling Applications Protection behaviour in terms of *uninstalling applications* can be predicted from NORM. However, the model explains only 6% of the variance (cf. Table 7.7).

Custom Location Running a simple linear regression analysis, it turned out that using the location obfuscation method of *custom location* cannot be predicted from BENE (Model 1, Table 7.8). Predicting the usage of the method from PCAL results in a model that can explain 9% of the variance in the usage (Model 2, Table 7.8). To evaluate the relationship of the predictors with the usage of *custom location* among the participants that used the method, the participants that did not use it were removed from the analysis. Spearman's rank-order correlations were calculated for the predictors and *custom location* with the remaining participants (N = 36). Usage of *custom location* was found to correlate with PCAL—the perception that risks outweigh the benefits in the usage of LBS ($r_s = 0.49$, $p = 0.004$). A subsequent simple linear regression suggests that, among the participants who used the method of custom location, 19% of the variance in the usage of the method could be explained by PCAL (Model 3, Table 7.8).

Switching Location Off BENE, PCAL, NORM, WILL, and DTRT all correlated with *switching location off*. No problems with multicollinearity were detected. Simple regression analyses suggest that the best predictors for this kind of behaviour are BENE, which can explain 13% of the variance in *switching location off*, NORM

Table 7.9 Summary of linear regression analysis predicting protection behaviour in terms of switching location off

Model	Adjusted R^2	F	Predictor	B	SE B	β	t
1	0.13	7.63**	Constant	0.24	0.18		1.34
			BENE	−0.14	0.05	−0.38	−2.76**
2	0.04	3.04	Constant	−0.20	0.08		−2.44*
			PCAL	0.16	0.09	0.25	1.74
3	0.12	6.93*	Constant	0.16	0.16		1.03
			NORM	−0.15	0.06	−0.37	−2.63*
7	0.14	4.53*	Constant	0.30	0.18		1.61
			BENE	−0.09	0.07	−0.25	−1.41
			NORM	−0.08	0.07	−0.21	−1.17

*$p < 0.05$; **$p < 0.01$

(explaining 12% of the variance), and WILL (explaining 11% of the variance). However, running a multiple linear regression using the stepwise method, the total model turns out to be significant, but the predicting variables do not predict *switching location off* statistically significantly. These results suggest that there may be some problems due to the collinearity of the predictors even though this was not identified from the multicollinearity statistics. In such a case the predictors are correlated, and as a consequence of the redundancy, the precision of the estimated regression coefficients decreases as more predictors are added to the model [70] (Table 7.9).

Increased Awareness Through Privacy Nudges
The influence of increased behavioural awareness through privacy nudges on privacy beliefs and behaviour was evaluated. In total 57% of the participants received privacy nudges, while the remaining 43% belonged to the control group. The influence of privacy nudges was assessed by comparing whether there were differences between the two groups on RISK and BENE, PCAL, LVAL, and privacy behaviours. It was found that the group that received privacy nudges ($M = 3.58, SD = 1.05$) reported higher perceived risks (RISK) than the control group ($M = 2.84, SD = 1.44$); $t(46) = 2.04, p = 0.047$. No differences were found between the two groups based on the other measured variables.

To further analyse the impact of increased awareness, only the nudged participants were included in the subsequent analysis while excluding the control group. Among the nudged participants, the relationship between privacy attitudes and privacy behaviour was assessed. The Spearman's rank-order correlation was conducted for the predictors and the behavioural measures (cf. Table 7.10). Some of the results deviate from those reported in Table 7.3; the small sample size could also cause issues with the correlation, as the influence of possible outliers attenuates [64]. Also, dissimilar distributions have stronger effect on correlations when sample size is small [64].

Table 7.10 Correlation coefficients for the measures in the field study when only the nudged participants are included

	Installing applications	Sharing location	Uninstalling applications	Custom location	Location off
RISK					
BENE				−0.37	−0.55**
PCAL				0.34	0.42*
NORM	0.51**	0.30	0.42**		−0.42*
WILL		0.38*			−0.40*
DTRT					0.47*
VIOL					
KNOW			0.47**		0.25
LVAL					0.53**

The results that differ from those in the whole sample are italicized. $^*p < 0.05$; $^{**}p < 0.01$. For clarity, only significant correlations, and those that were significant in the whole sample but not among the nudged participants, are reported

Among the nudged participants KNOW had a positive correlation with *uninstalling applications*, and LVAL with *switching location off*. Such results were not identified in the whole sample.

7.1.3   Intermediate Discussion

This chapter presents a field study evaluating privacy behaviours and their antecedents. The main focus was on assessing privacy protection behaviour in terms of using location settings on a smartphone application, and the influence that various location privacy attitudes have on such behaviour. Also location disclosing behaviour in terms of installing applications, and selecting a setting to share location, was analysed. This section starts by discussing the relationships between background factors and location privacy attitudes as presented in Chap. 2. Then the behaviours are explored, and whether or not they could be predicted from the attitudes.

7.1.3.1   Impact of Distrust and Willingness to Disclose

To study the relationships that had not been discussed in previous chapters, the influence of the willingness to disclose information, and that of distrust, on privacy attitudes was examined. The measure for willingness to disclose originates from a measure for inherent privacy concern, and comprises the negatively coded items in the construct. Thus, in a sense, willingness to disclose could be considered lack of privacy concern. The results of the study reveal that willingness to disclose significantly predicts social norm, which measures the extent to which the opinions of peers are perceived as important. Willingness to disclose also influences privacy calculus, mediated by perceived benefits. These results could be interpreted as that users who are more willing to share personal information about themselves are more susceptible to be influenced by one's peers. They are also more likely to find benefits in usage of location-based services, and through that, they are more likely to feel that benefits outweigh risks in using such services. Interestingly, no connection between willingness to disclose information and risk perception was identified. Distrust, which comprises the negatively coded items from the construct measuring trust, seems to negatively influence perception of benefits in the usage of LBS, which mediates the relationship between distrust and privacy calculus. Decreased perception of benefits increases the probability of feeling that risks outweigh benefits in the usage of LBS. This privacy calculus in turn seems to influence privacy valuation; however, the pathway from distrust to privacy valuation could not be explicitly identified by the statistics. These results from analysing the relationships between the background factors and location privacy attitudes suggest that, for these parts, the structure between the two first layers in the taxonomy presented in Chap. 2 reasonably depicts the relationships between these factors.

7.1.3.2 Impact of Privacy Knowledge

In contrast to the findings in Sect. 6.2, privacy knowledge was found to be correlated with privacy calculus and with distrust. These three predictors—privacy calculus, distrust, as well as privacy knowledge—correlate negatively with intention to use LBS and positively with switching location off, however, they were found to be poor predictors for these types of privacy behaviours. Additionally, among the participants who received privacy nudges, stronger privacy knowledge correlates positively with uninstalling applications. This result suggests that increased awareness of one's privacy behaviour might activate the privacy knowledge and lead to privacy protection behaviour in terms of uninstalling applications; however, this interpretation is merely speculative, and requires further empirical assessment. No findings suggesting a correlational relationship between knowledge and privacy behaviours were evident in Study V (cf. Sect. 6.2). The discrepancy in these results could be at least partially explained by the differences in the demographics of the participants taking part in these two studies—a larger portion of the participants worked in the IT sector in Study VI (19%), than in Study V (8%). However, no differences were found in how well participants working in the IT sector responded to the knowledge questions in comparison to others, which does not support the speculation.

7.1.3.3 Location Privacy Valuation

Risks, privacy calculus, distrust, as well as prior privacy violations correlate positively with higher location privacy valuation. However, location privacy valuation seems to translate to privacy behaviour only in the case when the users are nudged about their privacy behaviour—among the nudged participants, a positive correlation was found between location privacy valuation and switching location off. This result suggests that users who value their privacy higher are more prone to react to increased awareness by switching location services off. Other relationships with location privacy valuation and privacy behaviours were not identified. However, as other interpretations made about the influence of privacy nudges on behaviour, this is a speculative finding, which should be investigated further.

7.1.3.4   Awareness Through Privacy Nudges

A study by Almuhimedi et al. found evidence of privacy nudges positively influencing users to reassess (95% of the participants), and to change their privacy settings (58% of the participants) [10]. While it is possible that in the field study reported in this chapter the nudges did motivate the participants to switch location settings off, a discrepancy remains with the study by Almuhimedi et al., within which more than half changed the settings as a consequence of the nudges. One explanation as to why such results were not found in this work could be in the framing of the nudges—in this field study, the nudges did not explicitly state that the information was collected from the location information shared by the participants themselves. The explicit influence of privacy nudges on increased awareness was not assessed in this study, and thus their effectiveness cannot be evaluated beyond the speculations. However, the finding points out that the wording of privacy nudges is very important. The importance of wording can be further affirmed by the testimonies of several participants stating that they had found the nudges uncomfortable. Thus, although using privacy nudges can positively influence users into more privacy-conscious choices, care needs to be taken in the wording to create nudges that are efficient, and at the same time do not cause discomfort.

7.1.3.5   Predicting Privacy Behaviour

Most of the setting changes on the Protect Location application seem to have occurred during the first couple of days of the study, which suggests that the participants initially selected some privacy settings, and then rarely went back to change anything. This issue was overcome by considering the first 3 days of the study as an acclimatization period, during which the participants would get familiarized with the new devices. During the study, in particular the method of location obfuscation was not frequently used. One possible explanation to why using the method of obfuscation was used only scarcely could be that the concept and its efficacy might have been more difficult to grasp. This explanation is, however, not in line with the findings by Brush et al. [32] who suggest that users are able to understand different obfuscation methods. The question of why the method was used only little remains a question for future work. Because of the sparsity of the data for that particular type of behaviour, it was left out of the analysis.

Various regression models were suggested to identify which factors would adequately predict privacy behaviour. Normative beliefs seem to be a good choice for a predictor, as, among the measures included in this study, it showed the best performance for predicting most types of behaviours. However, normative beliefs were found not to influence intention to use LBS, or usage of location obfuscation through selecting a custom location. Presumably peers' influence can manifest in different ways in the context of protective behaviour—users to whom peers' influence is important are likely, to some extent, to adopt behaviours from their peers, irrespective of whether the peers' example would encourage them to adopt, or to neglect protective behaviours. The results from this study, however, indicate a positive relationship between normative beliefs and protective behaviours in terms of uninstalling applications, and switching location sharing off. Thus, it seems likely that the users who are more prone to be influenced by others also take protecting their privacy more seriously.

Privacy calculus and benefit perception were also identified as good predictors of privacy behaviour. Based on their interrelationships [48], as also identified in this work, benefit perception could be considered to be strongly integrated in privacy calculus. Thus, it should suffice to measure only privacy calculus—and indeed, the results from this study do not suggest that measuring perceived benefits would bring any added predictive power to the model.

Based on the analysis of the relationships between the measured variables representing background factors, and privacy attitudes and beliefs, it seems like a reasonable assumption to suggest that they follow the layered structure proposed in the taxonomy presented in Chap. 2. Then, based on this assumption, it should suffice to analyse the influence of the attitudinal variables for predicting privacy behaviours. However, the intention to use LBS is best predicted by willingness to disclose. This is a result that deviates from other findings in that other behavioural measures can be explained by attitudinal factors as also postulated in the taxonomy. This finding arises two further questions that remain topics for future research. First, should willingness to disclose be considered a privacy attitude rather than a background factor? Second, should intention to use LBS be considered a behavioural measure? This question arises also from the finding that intention in this study had only limited relationship with other behavioural measures.

7.2   Chapter Summary

In this chapter, usage of location protection mechanisms was evaluated during a field study, within which participants were provided a smartphone with a settings application, through which location sharing settings could be adjusted for each application on the device. While location obfuscation with a chosen accuracy was used only scarcely, the method of selecting a custom location, uninstalling applications, and switching location off were used relatively often.

There does not seem to be a single measure that would have equally strong influence on all these types of privacy behaviour. However, to a certain extent, all behavioural variables can be explained either by *social norm* (NORM), by *privacy calculus* (PCAL), or their combination. This finding seems to suggest that there are two main reasons driving users' behaviour in the context of location privacy: (1) the opinions of one's peers and (2) the assessment of whether benefits of using a service outweigh the possible risks.

Chapter 8
Discussion

8.1 Summary of Results

A set of research questions were posed along the lines of the LPT to frame the research done in this work. The following section summarizes the main results of the work in the framework of the posed research questions. The results are summarized following the layered structure of the LPT, starting with the background factors and how they are related to the location privacy attitudes and beliefs, and then moving on to predicting location privacy behaviours based on the attitudes and beliefs. For an illustration of the factors assessed within this work, cf. Fig. 8.1.

8.1.1 Background Factors Influencing Location Privacy Beliefs and Attitudes

This section discusses the empirical results in the light of the research questions posed in Sect. 1.3.1. These research questions consider the background factors influencing privacy attitudes in the context of location privacy; while these factors might be correlated with behaviours, a mediating relationship between these background factors and privacy behaviours through the privacy attitudes is proposed.

RQ 1.1. Privacy Concern
The first research question considers users' privacy concern, and how they influence the privacy beliefs and attitudes in the context of location privacy. Privacy concern is in this work considered as a trait, which is a rather stable part of users' personality, and could be considered to be the propensity to perceive privacy risks. Privacy concerns were first assessed in a qualitative study (cf. Sect. 4.1), which revealed the prevalence of the "nothing to hide" attitude—of users considering privacy as

© Springer Nature Switzerland AG 2020
M. E. Poikela, *Perceived Privacy in Location-Based Mobile System*,
T-Labs Series in Telecommunication Services,
https://doi.org/10.1007/978-3-030-34171-8_8

secrecy, and having a view that because they have done nothing wrong, they have nothing to hide [151]. This view is problematic, as it depreciates the importance of privacy, and, when getting more popularity, could lead to corrosion of privacy. In a field study assessing location disclosure in a context of location-based mobile participation, privacy concern was identified as a likely inhibiting factor for adoption of location-based services (Sect. 5.1), as also reported earlier by Xu et al. [176].

The influence of privacy concern—or the lack of it—was assessed quantitatively using a measure for dispositional privacy concern. However, when analysing the statistical properties of the construct, the six original items did not seem to be measuring the same phenomenon, and as a consequence, only the negatively coded items—measuring willingness to disclose personal information—were included in the measure. The construct is intended to measure the lack of privacy concern; however, it can be argued that what it measures is not exactly that, as in that case the items should have loaded on one factor within the EFA. Willingness to disclose was measured within a field study (cf. Sect. 7.1); it turned out to significantly predict subjective norm, and also privacy calculus, mediated by perceived benefits.

RQ 1.2. Prior Privacy Violations
Privacy violations that users have experienced in the past were studied within an online study (Sect. 6.2). Nearly all participants stated having experienced some privacy violations, and three quarters stated having received adverts based on their location; the received adverts were also seen predominantly discomforting. An exhaustive report of different kinds of privacy violations experienced was not done, and thus it is inconclusive which are the location privacy issues that are considered the most impactful; this remains a subject for future research.

In earlier works, experiencing privacy violations have been reported to affect privacy concern [150], and in online social networks, cause an individual to tighten their privacy protection mechanisms [156]. This work suggests that privacy violations have an impact on privacy attitudes also in the context of location privacy: experienced privacy violations were found to influence the extent to which users consider risks of LBS to outweigh the benefits, mediated by perceived risks.

RQ 1.3. Trust
One-third of interview participants stated trusting feelings, in particular towards big companies (Sect. 4.1). The comments expressed opinions that the companies have an "obligation" to handle users' data correctly, and not doing so would be harmful to their reputation. Trust was not identified as an influencing factor for disclosure in mobile e-participation (cf. Sect. 5.1). However, the influence of trust was assessed further in a crowdsourcing study, presented in Sect. 6.1, within which it was found to have an impact on likelihood that a location is shared: location sharing tasks were accepted less frequently when an untrusted advertiser was involved. Together with payment, trust could explain a significant proportion of location sharing when advertisers were present.

Within a field study (Sect. 7.1), distrust seems to influence privacy calculus mediated by perceived benefits, which is in line with the interrelationships postulated in the LPT.

RQ 1.4. LBS Knowledge

In an interview study presented in Sect. 4.1, several participants stated that they have knowledge gaps with respect to what happens to their location information once it has been shared; others had comments reflecting misunderstandings. It could be postulated that when users do not understand very clearly what happens with their data, they might also see their location information as less privacy sensitive and give it less value. Knowledge gaps could also lead these users to be more vulnerable to privacy attacks. Location privacy knowledge was explicitly measured within Study V (cf. Sect. 6.2), as well as Study VI (cf. Sect. 7.1) using a construct developed for assessing mobile users' knowledge with respect to location privacy. However, no relationship between location privacy knowledge and any of the measured privacy attitudes and beliefs was identified within the Study V. Within the Study VI location privacy knowledge was found to correlate with privacy calculus; however, it did not turn out to significantly contribute in predicting privacy calculus. Furthermore, correlations were identified between knowledge and behavioural constructs, but the relationships could not be verified with regression models.

Thus, this work did not succeed in clarifying whether or not being knowledgeable with respect to location privacy influences risk perception, location privacy valuation, and through these factors, privacy behaviour.

RQ 1.5. Demographics

Within Study V (cf. Sect. 6.2), the influence of demographic factors was found to influence beliefs and attitudes in the context of location privacy. Higher education was found to correlate with perceiving more risks in the usage of LBS. It was also found that women find LBS more often risky than men do, and are more prone to view risks to outweigh the benefits in the usage of LBS. These findings are in line with earlier literature [74, 146], and are a confirmation that the gender differences identified in perceived online privacy extend to the mobile context. The gender differences might stem from vulnerability in terms of physical safety—women perhaps consider themselves at a higher risk for violations such as stalking—however, this is merely a speculative argument and requires more work to assess.

RQ 1.5. Social Context

The influence of recipient of location information on location disclosure was studied within a field study in Sect. 5.2. No explicit relationship between a recipient of location information and location disclosure was identified. However, location was disclosed at a higher accuracy in situations in which the location requests were perceived as more pleasant. Additionally, the requests were considered as more pleasant when they came from individuals with whom the participants felt emotionally close.

As mentioned earlier, the recipient of location information had an impact within a crowdsourcing study (cf. Sect. 6.1)—according to the results, it seems that trust towards the recipient is an important factor in location sharing.

Social context is expected to influence behaviour mediated by social norm; however, this relationship is not explicitly studied in this work.

8.1.2   Location Privacy Beliefs and Attitudes Influencing Behavioural Outcomes

As discussed above, the interrelationships between the background factors and the factors constituting the beliefs and attitudes seem to be mostly in line with the layered structure of the LPT taxonomy. Thus, the background factors are postulated to influence behaviours through the beliefs and attitudes, and following the taxonomy, for predicting behaviours, it should therefore suffice to consider the beliefs and attitudes. These relationships are discussed here, following the research questions posed in Sect. 1.3.2.

RQ 2.1. Benefits and Risks
Various benefits in the usage of LBS were reported within an interview presented in Sect. 4.1. The most important benefit was reported as navigation, followed by saving time and effort through requiring less user input and thus simplifying interaction. Also social benefits were listed, including setting up meetings, and getting social recognition through socially desirable location reports to others. However, social recognition was always mentioned in a passive voice as something that others would do, but not the person reporting. Location information was also found useful for receiving personalized services. Other reported benefits included location-based games, sports, or collecting location information as a memento.

The most important reason for using a location-based mobile participation application was reported as the desire to help others, followed by getting stimulus against boredom, finding the topic (of a poll) important, as well as receiving a benefit (cf. Sect. 5.1).

The most frequently mentioned privacy risks associated with usage of LBS reported in the interview study included surveillance by the state or police, inappropriate use as well as secondary use of data, and user profiling. The risk of user profiling also included adverts based on location, which seems to polarize opinions between whether it is considered a benefit or a nuisance. The other risks specified by the participants were stalking and theft. Sometimes concerns regarding other, general privacy violations were voiced without an explicit clarification of what these might include.

RQ 2.2. Privacy Calculus Constituents
According to the theory of privacy calculus, users analyse the expected benefits of engaging in a behaviour and compare them with the perceived risks therein [48]. This calculation is then used as a guideline when determining whether or not to engage in the behaviour in question. This definition of privacy calculus is in line with the findings from this work: the perceived risks and the benefits were found to be influencing factors for privacy calculus also in the context of location privacy (cf. Sect. 6.2). Half of the participants within an interview study stated that using location-based services is a trade-off, within which they accept some privacy risk, in order to receive a desirable service (Sect. 4.1).

RQ 2.3. Privacy Calculus

Perception of risks was in a qualitative study connected with using more protective measures, such as switching location services off or restricting location sharing, avoiding usage of LBS, or educating oneself (Sect. 4.1). Also evidence of risk perception influencing location disclosure was found: in Study III a correlation between risk perception and location sharing accuracy was identified (cf. Sect. 5.2). These results were, however, not confirmed within a field study (Sect. 7.1).

The results from this work suggest that perception of benefits in the usage of LBS and privacy calculus are rather good predictors of privacy behaviour. Though, as previously stated, perceived benefits and perceived risks are the constituents of privacy calculus, which leads to the conclusion that inclusion of risks as well as benefits can be superseded by privacy calculus alone. This statement is challenged by the fact that for some types of behaviour perceived benefits alone worked as a stronger predictor than privacy calculus alone. Privacy calculus was found to predict location disclosure behaviours in terms of intention to use LBS, as well as location privacy protection behaviours, in particular the usage of obfuscation.

The single-indicator variable of privacy calculus was not included in the measurement model (cf. Chap. 3), but as it seems to be of high importance in predicting privacy behaviour in the context of location, it should be introduced and its influence systematically assessed.

RQ 2.4. Social Norm

A positive correlation was identified between social norm and self-reported usage of LBS (cf. Sect. 6.2); however, the contribution of social norm in predicting this kind of behaviour was not significant.

Based on the results from a field study, presented in Sect. 7.1, it seems that social norm has a major role in defining privacy behaviour—positive correlations were found with disclosing behaviours in terms of installing applications and sharing location, and negative correlations with protection behaviours in terms of uninstalling applications, and switching location sharing off. In all these cases, social norm was also contributing significantly in predicting the behavioural outcome. This could suggest that influence of others' expectations on one's privacy behaviour might even surpass the effect of the privacy calculus. This finding is not conclusive and would require further studies to be confirmed.

RQ 2.5. Perceived Control

As the efforts in developing a measure for perceived control turned out unsuccessful, systematic assessment of perceived control and its relationship with other variables in the model remains a subject for future research.

RQ 2.6. Location Privacy Valuation

Three methods were used for quantifying location privacy through monetary valuation, which was used to assess whether or not this valuation can provide a predictor for privacy behaviour in the context of location. First, in a crowdsourcing study presented in Sect. 6.1, the participants were offered a small monetary compensation for sharing their current location. Then, in an online study presented in Sect. 6.2, the participants could define how much they would need to be compensated for giving away their location data for the following 2 weeks. Finally, in a questionnaire administered after a 15-day field study (cf. Sect. 7.1), the participants could define how much they would need to be compensated for giving away the location data that has been collected of them during the past 2 weeks.

To assess the relationship between privacy calculus and location privacy valuation, these variables were measured within the online study. The results from the study show that privacy calculus correlates positively with higher location privacy valuation; however, a regression analysis showed that only a small portion of the variance in location privacy valuation could be explained by the privacy calculus. However, repeating the same analysis within the field study reveals that privacy calculus can explain 16% of the variance in location privacy valuation. This discrepancy could be explained by the difference in the assessment of location valuation in these two studies—attributing monetary value to the location data collected in the future is a more hypothetical scenario, as the participant might not yet know which locations they would visit. On the other hand, considering the valuation retrospectively gives the participant a chance to estimate more concretely how privacy infringing it would be to disclose the location information. Thus, based on these findings, it can be concluded that location valuation can to some extent be considered to quantify privacy calculus, when the user has a chance to (retrospectively) evaluate the location disclosure.

How location privacy valuation translates to privacy behaviour was evaluated within the aforementioned three studies. The findings from the online study did not imply a relationship between location privacy valuation and usage frequency of LBS, which could be a direct consequence of the issue discussed above. The results from the field study suggest that location privacy valuation translates to privacy behaviour only in the case when the users are nudged about their privacy behaviour—among the nudged participants, a positive correlation was found between location privacy valuation and switching location off. Within the crowdsourcing study, however, the provided compensation was found to be a significant contributor to likelihood of sharing a current location.

As a conclusion, monetary quantification seems to be a rather effective method for estimating location disclosing behaviour in a one-time sharing scenario in-situ. However, the prediction power is not as evident in the case when a compensation is provided for location traces collected in the past. This work did not find evidence that the method would be effective in the case that the location information has not yet been collected.

8.1.3 Behavioural Outcomes

RQ 3.1. Intention
Intention to use LBS was best explained by willingness to disclose personal information, rather than by factors categorized as beliefs and attitudes. This finding

deviates from the results with respect to other behavioural outcomes, for which the beliefs and attitudes worked as reasonably good predictors.

Intention to use LBS was found to have a positive correlation with installing applications; however, it did not significantly contribute to the behaviour as the variance explained by intention was rather low. No relationship was identified between intention and other behaviours; however, this finding is little surprising considering that the intention to use LBS was measured, rather than the intention to use privacy protection mechanisms.

RQ 3.2. Awareness

The influence of increased awareness of one's privacy behaviour was assessed within a field study, presented in Sect. 7.1, through privacy nudges. The results from the study suggest that the privacy nudges increased risk perception in participants, as the nudged group reported higher risk perception scores than the control group. There was, however, no clear difference found between the two groups in privacy behaviours. Among the nudged participants location privacy knowledge was found to have a positive correlation with uninstalling applications, and similarly, location privacy valuation was found to have a positive correlation with switching location off. These relationships were not found in the control group that did not receive nudges. The results could be interpreted as that increased awareness has the effect that users with stronger knowledge with respect to location privacy protect themselves by uninstalling applications, and on the other hand, those who attribute higher monetary value to their location privacy switch location services off. However, as there was no measurable difference between the experimental group and the control group, more work is required to systematically assess the relationship.

RQ 3.3. Behavioural Model

Location privacy was in this work analysed and estimated in the framework of a taxonomy, which presents privacy behaviours, location privacy beliefs and attitudes, as well as the related background factors in three separate, interconnected layers. Throughout this work these factors were assessed, and their interrelationships were estimated using statistical methods. The studied factors and their interrelationships are illustrated in Fig. 8.1.

The first layer of the taxonomy, *Background Factors*, was assessed mainly with respect to the user factors, including the privacy concern (or the lack of it), prior privacy violations, trust and distrust, as well as location privacy knowledge, and demographics. The influence of personality was not extensively assessed within this work. Also, the system factors were not considered, and with respect to the context factors, mainly the social context was assessed. The factors that were assessed within this layer were found to have a relationship with factors within the second layer predominantly consistently with the taxonomy.

The second layer, *Beliefs and Attitudes*, presumably plays a major role in predicting privacy behaviours. Therefore, effort was put into creating a measurement model comprising of the variables within this layer. The model could however not be validated. The problem resulting from using unvalidated questionnaires,

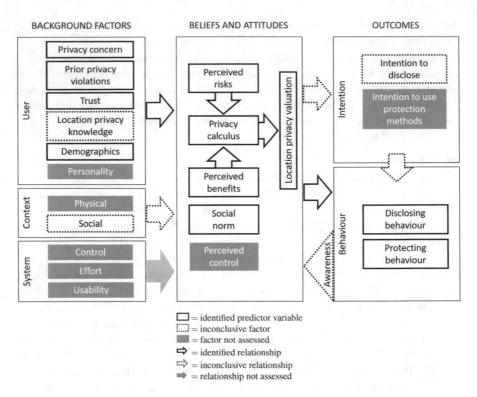

Fig. 8.1 The factors as well as relationships postulated in the LPT. The factors depicted in grey boxes have not been operationalized successfully in this work. The factors depicted with boxes with dotted outlines have been operationalized, but were not shown to contribute in predicting the factors in the subsequent layer. The factors depicted with white boxes and solid black lines were found to predict the factors in the subsequent layer. The grey arrow signifies a relationship that was not assessed in this work. The dotted arrows depict relationships that were studied, but require more work to reach sufficient conclusions. The white arrows with solid black lines depict relationships that were identified in this work

which do not necessarily display the required measurement properties, is that a risk of measurement error persists, and therefore the results drawn from using such questionnaires must be interpreted with care. Thus, lamentably, the conclusions derived from the empirical studies cannot be stated with full confidence.

Intention to use protection methods was not assessed within this work. Intention to disclose location was assessed in terms of intention to use location-based services. In this work, intention could not be predicted by beliefs and attitudes, but rather by willingness to disclose personal information, which is considered to be a trait, and categorized as a background factor. This finding could suggest that the structure of the taxonomy is not optimal for the behavioural intention, but, assuming that the taxonomy otherwise has a correct structure, intention could rather be considered more as an attitudinal factor, which is also directly influenced by background factors. Another solution to this could be that the assumption that privacy concern and the lack of it—willingness to disclose personal information—behave rather as states than traits, and should be treated as factors within the layer for beliefs and attitudes.

Privacy behaviour on online SNSs has in earlier works been explained in the light of the TPB from privacy concerns, attitudes, and intentions [46]. In this work the relationship between behavioural intention and behaviour could be identified only partly—more work is needed to draw final conclusions on whether or not the relationship exists. Privacy behaviours could be reasonably predicted from social norm, perceived benefits, and privacy calculus. Perceived risks are included in the prediction through privacy calculus, and therefore a lack of evidence of direct influence does not mean that an influence does not exist.

8.2 Challenges

The studies in this work relied on recruitment methods that could be susceptible to self-selection bias, and as a consequence, the most privacy aware users might not participate in this kind of studies. The issue is, however, difficult to overcome, and would require either studying users of existing services or conducting surveys using systematic sampling methods other than those based on self-selection. Both methods have drawbacks. First, the user base of existing services might not be representative of the whole population; for example, the most privacy-concerned users might not be as likely to adopt social networking services, or location-based services [176]. Moreover, studying usage of existing applications might limit the experimental design due to limited possibility to log user data, as well as limited control options provided by the applications. The most obvious problems with survey studies are the biases inherent in self-reporting. The responses might suffer from social desirability bias [115], and also from errors arising because of issues with recalling when reporting one's past behaviour.

A participants' database was used in most of the studies to recruit all or some of the participants, and in some cases, also to handle scheduling of meetings. All participants had to register to the database in order to participate, which required giving out personal information, including full name, email address, and phone number. This requirement might have had an influence on the willingness of some potential participants to take part in the studies. The most likely result from this is an under-representation of the most privacy-concerned individuals in the sample.

Social facilitation might play a role in some of the findings—certain individuals are more susceptible to the social influence than others [107]. Evidence to support this can be seen in that the social norm was found to have a relationship with privacy behaviour: it is likely that participants with higher scores on the social norm scale are more susceptible to alter their behaviour according to what they think is expected of them. Such altering of behaviour could happen as a consequence of an observer effect due to the usage of LBS, or even as a consequence of participation in a study. It cannot be stated with certainty whether one of these phenomena took place. How to overcome the uncertainty is non-trivial, as in order to conduct studies—or any data collection conforming to ethical standards—the participants need to explicitly agree to the collection of data, which, consequently, makes them susceptible to the observer effect.

8.3   Methodological Recommendations for Privacy Research

Several challenges are met when studying users' privacy attitudes and behaviour. First, mentioning privacy causes users to be tuned to think about privacy, and therefore to state stronger privacy attitudes. Thus, care should be taken in wording of questionnaire items not to influence the users' ratings towards higher privacy concern. Also, surveys might altogether not be the best way to gain understanding of users' privacy behaviour, as it might not be in line with the stated attitudes [2, 153]. This is why it is important to study privacy behaviour in real-life scenarios. One option to overcome this challenge, as discussed earlier, is to study users of existing applications; however, this limits the possibilities to control the experimental setting. Thus, the solution to overcome these issues in several studies presented in this work was to use deception to pose a hypothetical scenario. This was done to create an experimental setting within which the participants could react as genuinely as possible. There is evidence indicating that the hypothetical scenarios were believable, for example, participants stated emotional responses to having received location requests within Study III (cf. Sect. 5.2). Thus, the method can be recommended for future studies.

In this work, participants were offered test devices as a method to overcome the issue that the test application would not work on the participants' devices with full functionalities. The method required the users to switch to the test device as a main device, and some participants were somewhat unsure about how this would work for them. There were some issues experienced with the test devices, and some participants even cancelled their participation during the field study because of these issues. The overall experience from the perspective of running the study was, however, rather positive, as the method allowed a study to be run that would have otherwise required using an app without actual functionality for location obfuscation. The drawbacks from the method were mostly due to the test devices being of low quality—in the cases of unhappiness voiced by the participants, of lower quality than the ones the participants were used to. Thus, the method can be recommended for privacy research, on the condition that higher quality devices are used.

Most importantly, as the results from the Studies IV (Sect. 6.1), V (Sect. 6.2), and VI (Sect. 7.1) reveal, privacy calculus and location valuation are very promising factors that, when used properly, can provide a tool to assess location privacy behaviour without the need to administer large questionnaires, which consume resources of the participants and researchers alike.

8.4   Future Work

Other factors that might influence privacy attitudes and behaviour in this context may include user-specific, culture-specific, environmental, context-related, as well as system-related factors. A lot of work needs to be done to cover these aspects and their influence on privacy behaviour in the context of location sharing.

Privacy violations experienced by users were assessed in this work within an online study. However, the list of experienced violations acquired from the study is not exhaustive, and future research should address this topic.

The risks arising from not using a service are not assessed in this work. These could include social ostracizing when a service that the peers use actively is not adopted. Such risks could be partly covered within the factor for social norm; however, they may not be limited to social risks.

Privacy was in this work defined as "the selective control of access to the self" [12]. In the context of information privacy this control refers to the control that an individual has over how their personal data is collected and used [174]. Following these definitions and relying on the theory of planned behaviour, the perceived control was included in the postulated taxonomy as an antecedent to privacy behaviour. However, the construct did not turn out satisfactory, and was therefore not considered further. The inclusion of perceived control in the taxonomy should be addressed in future studies.

Sometimes sharing a semantic location would be more meaningful than the physical coordinates. In such situations the user can of course send a message in free-text form, however, also other solutions could be explored. In this study the users could select a semantic location from a drop-down menu; however, it would be more applicable from the user's perspective to provide a possibility for the user to modify the semantic sharing options.

In this study, intention to use LBS was assessed as a predictor for privacy behaviour in terms of installing applications and sharing location information. For evaluating the influence of intention on protection behaviours, intention to protect should be explicitly measured. This remains a topic for future research.

Privacy nudges were used to increase participants' awareness about how they are sharing their location information. Some participants' comments in the debriefing suggest that the nudges might have caused discomfort. While it can be considered meaningful to attempt to increase participants' awareness of their location disclosures, it is also important to not cause any discomfort. How the nudges influence participants beyond the altered privacy behaviour should be assessed in future research.

Finally, user's physical location can be considered a particularly privacy sensitive piece of information. However, also other types of sensory data could be considered, and the adequacy of the model assessed in other contexts. This would require the constructs to be adapted to the new context, and their interrelationships systematically assessed.

Appendix A

A.1 Study I: Interview Script

- How many location-sharing applications do you have on your smartphone?

  - Which ones?

- Which other applications do you have?

  - Are there some applications that potentially use location features without your knowledge?

- Why are the mentioned location-based applications being used?
- If you do not use location-based applications, why not?
- What are some possible benefits you think there are in using location-based applications?
- What kind of benefits have you already had?
- Have you heard of any possible risks that there might be?

  - What risks?
  - How did you heard about the risks?

- How has the knowledge of possible risks affected the use of location-based applications?

  - Have you chosen not to install some applications?
  - Have you used applications less or differently because of the knowledge?

- What do you think is done with your data?
- Do you believe the companies that create location-based applications can access your location data?
- What do you believe the companies do with the location data?
- What do you believe is possible to do with the location data?

© Springer Nature Switzerland AG 2020
M. E. Poikela, *Perceived Privacy in Location-Based Mobile System*,
T-Labs Series in Telecommunication Services,
https://doi.org/10.1007/978-3-030-34171-8

- How likely do you believe it might be that ...

    - ... your home or work address becomes known?
    - ... the data is collected to be sold to third parties such as advertisers?
    - ... the data is collected to create a profile of you?
    - ... the data is combined with other information to create a profile of the user?
    - ... and sold to advertisers?

A.2 Study II: List of All Polls

A complete list of all polls distributed within the Study II is listed in Table A.1. Also a translation from the original German text into is provided.

A.3 Study VII: Questionnaire Construction

This study ($N = 39$) was conducted as a closing questionnaire of a field study with all German participants [67], for which the participants received a monetary incentive of 8–15€ based on their activity in participating. The participants were recruited mainly through a participants' database Prometei offered by TU Berlin, but also through online classified adverts service called Ebay Kleinanzeigen,[1] and physical billboard posters. The questionnaire was administered online using the LimeSurvey platform in July 2015, and took roughly 20 min to complete. The scales deployed in this questionnaire were: PUSU, UNAUT, ACCU, COLL, RISK, BENE; NORM, and DFP (WILL).

Of the initial 40 participants, one was excluded because of the evident non-committal in responding, and all analyses are done with the remaining 39 participants. The participants' mean age was 29.5 yrs; the other demographic details are presented in Table A.2.

A.4 Study VIII: Questionnaire Construction

An online study was conducted using the Limesurvey Platform with the aim of validating the risk perception questionnaire ($N = 34$). The participants were recruited through the Prometei portal, and were compensated with a 5€ Amazon voucher. The scales deployed in this study were PUSU, UNAUT, ACCU, COLL, RISK, and DFP (WILL).

---

[1]http://ebay-kleinanzeigen.de/.

Table A.1 List of all eight polls distributed during the Study II, including the original German versions, as well as translations into English

	Original German wording	English translation
Poll 1		
Title	Laptop, Tablet, oder Smartphone?	Laptop, tablet, or smartphone?
Subtitle	Verdienen Sie 2€ mit der Teilnahme an einer kurzen Studie (2 min.) von Bachelorstudenten der TU Berlin!	Get a 2€ voucher by participating in this short (2 min) Bachelor study, TU Berlin
Intro	Die folgende Umfrage ist Teil eines Bachelor-Studienprojekt der TU Berlin. Das Studienprojekt beschäftigt sich mit der Frage welche Endgeräte—PC, Tablet oder Smartphone, bei der Internetnutzung bevorzugt werden. Für Ihre Teilnahme werden sie mit einem Amazon Gutschein über 2€ belohnt—um den Gutschein zu erhalten, geben Sie bitte Ihre E-Mailadresse an! Sämtliche von Ihnen zur Verfügung gestellten Informationen werden ausschließlich für wissenschaftliche Zwecke genutzt und nicht an Dritte weitergegeben	This questionnaire is conducted as part of a bachelors' study project at TU Berlin, focusing on the question of what the preferred method of using the internet is—via a PC, tablet, or a smartphone. For participating, you will be rewarded with a small voucher (2€)—please give your email address for receiving the voucher via email! All the information that you provide will be used strictly for scientific purposes only and not given to any third parties
Q1	Welche/-s Endgerät/-e nutzen Sie um im Internet zu surfen? Bitte wählen Sie alle zutreffenden Antworten aus	Which device(s) do you use for browsing the internet? Please mark all that apply
	– Desktop Computer	– Desktop computer
	– Laptop	– Laptop
	– Tablet	– Tablet
	– Smartphone	– Smartphone
	– Fernseher mit Internetanschluss	– Internet-enabled TV
	– Spielkonsolen	– Games machine
	– Andere	– Other
Q2	Warum nutzen Sie diese/-s Gerät/-e zum Surfen im Internet? Bitte wählen Sie alle zutreffenden Antworten aus	Why do you use the device(s) that you do for browsing the internet? Please mark all that apply
	– Ich besitze keine anderen Geräte	– I have no other devices
	– Ich finde diese/-s Gerät/-e am praktischsten	– I find it most practical to use
	– Ich wechsle zwischen den Geräten je nach Kontext (auf der Arbeit, Zuhause/ unterwegs)	– I change between devices based on the context (at work or school/at home/on the way)
	– Andere	– Other
Q3	Wie lange sind Sie während der letzten zwei Wochen pro Tag durchschnittlich im Internet gesurft? Bitte geben Sie eine Schätzung in Stunden und Minuten an	During the past 2 weeks, how much did you use internet on average per day? Please give an estimate in minutes or hours

(continued)

Table A.1 (continued)

	Original German wording	English translation
Q4	Wie viele Personen, Sie selbst eingeschlossen, leben ständig in Ihrem Haushalt?	How many people, including yourself, live in your household?
Q5	Sind sie männlich oder weiblich?	Are you male or female?
	– Männlich	– Male
	– Weiblich	– Female
Q6	Bitte geben Sie Ihre E-Mailadresse an, damit wir Ihnen Ihren Gutschein zusenden können. Möglicherweise senden wir Ihnen auch eine Einladung zu einem weiteren Fragebogen an diese E-Mailadresse!	Please share your email address so that we can send you the gift voucher. We might also send you an invitation for a further questionnaire via this email address!
Poll 2		
Title	Über Ihre Smartphone-Nutzung	About Your Smartphone Usage
Subtitle	Für die Beantwortung der Umfrage erhalten sie einen 2 € Amazon Gutschein. Statistisches Bundesamt Destatis	Responses are credited with a 2€ Amazon voucher. Statistisches Bundesamt Destatis
Intro	Ziel dieser vom Statistischen Bundesamt Destatis durchgeführten Studie ist es, das Nutzungsverhalten von derzeitigen Smartphone-Nutzern zu untersuchen. Ihre Daten werden anonymisiert und vertraulich behandelt. Der kurze Fragebogen umfasst sieben Fragen. Die ersten drei Fragen befassen sich mit Smartphone-Nutzung, die übrigen vier Fragen erfassen demographische Merkmale. Für Ihre Teilnahme erhalten Sie einen Gutschein über 2 €. Um den Gutschein zu erhalten, klicken Sie auf den Link am Ende es Fragebogens	The goal of this study by Statistische Bundesamt Destatis is to study smartphone usage of current smartphone users. Your data is anonymized and kept confidential. This short questionnaire has seven questions in total. The first three questions focus on smartphone usage, followed by four demographic questions. You will receive a 2€ voucher as an incentive. To collect the voucher, please give your email address at the end of the poll
Q1	Seit wann besitzen Sie ein Smartphone?	How long have you been using a smartphone?
	– Weniger als 1 Jahr	– Less than 1 year
	– 1-3 Jahre	– 1–3 years
	– Mehr als drei Jahre	– More than 3 years
	– Ich besitze kein Smartphone	– I do not have a smartphone
Q2	Wie häufig haben sie im letzten Monat ihr Smartphone durchschnittlich genutzt?	On an average day during last month, how much did you use a smartphone?
	– Ich nutze das Smartphone normalerweise nicht	– I do not normally use a smartphone
	– Weniger als einmal pro Woche	– Less than weekly
	– Ein- bis mehrmals pro Woche	– Once to a few times a week
	– Ein- bis mehrmals pro Tag	– Once to a few times a day
	– Ein- bis mehrmals pro Stunde	– Once to a few times an hour

(continued)

Table A.1   (continued)

	Original German wording	English translation
Q3	Für welche der folgenden Aktivitäten haben sie ihr Smartphone bereits genutzt? Bitte wählen Sie alle zutreffenden Antworten aus	Have you ever used your smartphone for one or several of the following? Please mark all that apply
	– Fotos machen	– Taking a picture
	– Senden oder Empfangen von Textnachrichten	– Sending or receiving text messages
	– Spielen	– Playing a game
	– Senden oder empfangen von E-Mails	– Sending or receiving email
	– Im Internet surfen	– Accessing the internet
	– Abspielen von Musik	– Playing music
	– Senden oder empfangen von Chatnachrichten	– Sending or receiving instant messages
	– Aufnehmen eines Videos	– Recording a video
Q4	Wie alt sind Sie?	How old are you?
Q5	Was ist Ihr höchster Schulabschluss?	What is your highest education level?
	– Hauptschulabschluss	– Secondary school (Hauptschulabschluss)
	– Mittlere Reife	– Secondary school (Mittlere Reife)
	– Abitur (inkl. Fachabitur)	– Highschool
	– Abgeschlossenes Studium (Uni, FH)	– University or university of applied sciences
	– Kein Schulabschluss	– No education
Q6	Bitte teilen Sie uns Ihre E-Mailadresse mit, um Ihren Gutschein zu erhalten und um uns die Möglichkeit zu geben Sie zu weiteren Fragen zu kontaktieren.	Please share an email address for receiving a gift voucher and giving us a possibility to ask you further questions
Poll 3		
Title	Helfen sie mit bei der Verbesserung des öffentlichen Personennahverkehrs!	Get your public transportation network improved!
Subtitle	Berliner Verkehrsbetriebe BVG in Kooperation mit der Deutschen Bahn	Berliner Verkehrsbetriebe BVG in co-operation with Deutsche Bahn
Intro	Beantworten Sie die Umfrage um uns bei der Verbesserung unseres Services zu helfen! Außerdem erheben wir einige demografische Eigenschaften	Respond to this survey to help us improve our service! We would also like to ask you for some demographic details
Q1	Haben sie während der letzten sechs Monate die Angebote der BVG genutzt?	During the past 6 months, have you used some of the following of BVG's and Deutsche Bahn's services?
	– Ja, ich habe die Regionalbahn genutzt	– Yes, I have traveled by the local trains (Regionalbahn)
	– Ja, ich habe S-Bahn genutzt	– Yes, I have traveled by the S-Bahns
	– Ja, ich habe die U-Bahn genutzt	– Yes, I have traveled by the U-Bahns
	– Ja, ich habe die Tram genutzt	– Yes, I have traveled by the Trams

(continued)

Table A.1 (continued)

	Original German wording	English translation
	– Ja, ich habe den Bus genutzt	– Yes, I have traveled by the Buses
	– Ja, ich habe die Fähre genutzt	– Yes, I have traveled by the Ferries
	– Ja, ich habe ein Angebot der BVG genutzt, kann mich aber nicht mehr erinnern welches	– Yes, but I do not remember which type of transportation I used
	– Nein, ich habe während der letzten sechs Monate keines der BVG- Angebote genutzt	– No, I have not used any of these services during the past 6 months
	– Anderes [ ]	– Other [ ]
Q2	Wie ist Ihre Meinung zur kürzlichen Erhöhung der Fahrpreise?	What do you think about the recent increments in the ticket prices?
	– Die Preise waren bereits zu hoch und hätten eher gesenkt werden sollen	– The prices were already too high and should have rather been decreased
	– Die Preise waren angemessen und hätten nicht erhöht werden sollen	– The prices were already appropriate and should not have been increased
	– Die Fahrpreiserhöhung ist mir egal	– I don't mind the increment in the prices
	– Die Preise waren vorher zu niedrig; es ist richtig, dass sie erhöht wurden	– The prices were previously too low, it was fair to increase them
	– Ich habe nicht bemerkt, dass sich die Preise geändert haben	– I did not notice that the prices were changed.
	– Anderes [ ]	– Other [ ]
Q3	Wie zufrieden sind Sie mit dem Service der BVG?	Are you satisfied with the BVG's and Deutsche Bahn's services in and around Berlin?
	– Sehr unzufrieden	– Very unsatisfied
	– Unzufrieden	– Unsatisfied
	– Weder zufrieden noch unzufrieden	– Neither satisfied nor unsatisfied
	– Zufrieden	– Satisfied
	– Sehr zufrieden	– Very satisfied
Q4	Wie alt sind Sie?	How old are you?
	– Unter 18 Jahre	– Below 18 years
	– 18–24 Jahre	– 18–24 years
	– 25–40 Jahre	– 25–40 years
	– 41–55 Jahre	– 41–55 years
	– 56–65 Jahre	– 56–65 years
	– 66 Jahre und älter	– 66 years and older
Q5	Was ist Ihr höchster Schulabschluss?	What is your highest education level?
	– Hauptschulabschluss	– Secondary school (Hauptschulabschluss)
	– Mittlere Reife	– Secondary school (Mittlere Reife)
	– Abitur (inkl. Fachabitur)	– Highschool
	– Abgeschlossenes Studium (Uni, FH)	– University or university of applied sciences
	– Kein Schulabschluss	– No education

(continued)

Table A.1   (continued)

	Original German wording	English translation
Q6	Möchten Sie, dass wir sie zu Ihrem Feedback kontaktieren? Bitte hinterlassen Sie Ihre E-Mailadresse damit wir sie erreichen können	Would you like us to contact you regarding your feedback? Please leave your email address so that we can get back to you!
Poll 4		
Title	Sind Sie mit Ihrer Netzabdeckung zufrieden? Helfen Sie sie zu verbessern!	Are you satisfied with your network coverage? Help to enhance it!
Subtitle	Kooperation mit Telekom, Vodafone, E-Plus und O2	Collaboration of Telekom, Vodafone, E-Plus, and O2
Intro	Sehr geehrte Mobilfunknutzer, um die Netzqualität zu verbessern, führen wir nicht nur Messungen und Analysen zur Netzabdeckung, sondern auch Kundenzufriedenheitsbefragungen durch, um mehr über die subjektive Qualität unserer Netzabdeckung zu erfahren. Anhand dieser Umfragen, streben wir eine Verbesserung des Service für Sie an. Diese Studie wird in Kooperation mit Telekom, Vodafone, E-Plus und O2 durchgeführt	Dear mobile phone user, in an attempt to improve the teleservices in Germany, we are conducting not only measurements and network coverage analysis, but also customer satisfaction surveys to learn about the subjective quality of our coverage. Based on these surveys, we aim at enhancing the service for you. This study is done in collaboration with Telekom, Vodafone, E-Plus and O2
Q1	Welche Mobilfunkanbieter nutzen sie momentan?	Which mobile operator do you currently use?
	– Telekom	– Telekom
	– Vodafone	– Vodafone
	– E-Plus	– E-Plus
	– O2	– O2
	– Andere [ ]	– Other [ ]
Q2	Wieviele MINUTEN pro Tag nutzen Sie ihr Mobiltelefon ungefähr für Sprachanrufe?	How many minutes do you approximately use your mobile phone for making phone calls each day?
Q3	Wie zufrieden sind sie mit Ihrer Netzabdeckung innerhalb von Deutschland?	How satisfied are you on your network coverage inside Germany?
	– Sehr unzufrieden	– Not at all satisfied
	– Unzufrieden	– Slightly satisfied
	– Weder zufrieden noch unzufrieden	– Moderately satisfied
	– Zufrieden	– Very satisfied
	– Sehr zufrieden	– Extremely satisfied
Q4	Sind sie männlich oder weiblich?	Are you male or female?
	– Männlich	– Male
	– Weiblich	– Female
	– Keine Angabe	– I prefer not to say

(continued)

Table A.1 (continued)

	Original German wording	English translation
Q5	Welche der folgenden Aussagen trifft auf Ihren derzeitigen beruflichen Status bzw. Ihre Ausbildung zu?	Which of the following is true regarding your current employment status and your education?
	– Voll berufstätig	– Fulltime employed
	– Teilweise berufstätig Teilzeit/stundenweise/zeitweise	– Partial time employed—part time/hourly/temporary
	– Vorübergehend nicht berufstätig, arbeitslos	– Temporarily not employed, unemployed
	– Nicht mehr berufstätig—in Rente/Pension	– No longer working—in retirement/pension
	– Nicht (mehr) berufstätig—Hausfrau/Hausmann	– No (longer) working—housewife/homemaker
	– In Berufsausbil-dung/Lehre/Wehrpflicht/Zivildienst	– In occupational education/teaching/military/civil service
	– In Schulausbildung – Schüler	– In basic education—pupils and students
	– In Hochschulausbildung—Student	– In higher education–students
	– Nicht berufstätig und nie berufstätig gewesen	– Not employed and never worked
Q6	Unter Umständen haben wir weitere Fragen zu Ihrem Feedback und würden Sie in diesem Fall gerne kontaktieren. Wenn Sie dem zustimmen, hinterlassen Sie uns bitte Ihre E-Mailadresse.	We might have some further questions regarding your feedback, and would need to contact you for this. If you agree with us contacting you, please share your email address!
Poll 5		
Title	FlashPoll	FlashPoll
Subtitle	Eine kurze Umfrage zur Optimierung der FlashPoll App	A short poll to improve the FlashPoll app
Intro	Flashpoll möchte eine erfolgreiche Platform zur mobilen Partizipation sein. Um dieses Ziel zu erreichen würden wir gerne wissen, wie wir uns verbessern können. Außerdem möchte wir gerne wissen, wer unsere Nutzer sind. Dazu möchten wir Sie bitten, die folgenden Fragen zu FlashPoll und zu Ihrer Person zu beantworten	FlashPoll aims at being a successful mobile participation platform. To reach our goal, we would like to learn how we could improve. Also, we would like to know who our users are. Please respond to a few questions regarding FlashPoll, and to a few regarding yourself
Q1	Was ist Ihrer Meinung der größte Vorteil von Flashpoll?	What is, in your opinion, the main advantage of FlashPoll?
	– Da die Umfragen ortsbasiert sind, werden die Bürger, für die die Befragung relevant ist, besser erreicht	– Localization permits the polls to be better targeted for relevant audience
	– Die Stadtverwaltungen können schneller Feedback Ihrer Bürger einholen	– The government can get faster feedback from the citizens

(continued)

Table A.1 (continued)

	Original German wording	English translation
	– Durch die Umfragen können Stadtverwaltungen besser mit Ihren Bürgern kommunizieren	– The polls allow better communication between the municipalities and the citizens
	– FlashPoll ist schnell und einfach zu benutzen	– FlashPoll is quick and easy to use
	– Anderes: [ ]	– Other
Q2	Was könnte Ihrer Meinung nach an Flashpoll verbessert werden?	In your opinion, how could FlashPoll still be improved?
Q3	Wie haben Sie von Flashpoll erfahren?	How did you hear about FlashPoll?
	– über einen Freund	– Through a friend
	– Durch Werbung	– From an advertisement
	– Bei einem Event	– In an event
	– Online oder im Appstore	– Online or app store
	– Anderes:	– Other
Q4	Welche Medien nutzen sie um sich zu informieren?	Which media do you typically use for informing yourself?
	– Nachrichtenseiten im Internet	– News sites on the internet
	– Andere Seiten im Internet	– Other sites on the internet
	– Zeitungen	– Newspapers
	– Radio	– Radio
	– TV	– TV
	– Anderes:	– Other
Q5	Was ist gegenwärtig Ihr höchster (Aus-) Bildungsabschluss?	What is your highest educational level?
Q6	Einige der Studienteilnehmer würden wir nach Abschluss der Studie ggf. kontaktieren um eventuell auftretende Rückfragen zu klären. Bitte geben Sie dazu Ihre E-Mail-Adresse an:	We would like to contact some of our users for some further information. Please give us your email address:
Poll 6		
Title	RTL Television	RTL Television
Subtitle	Nehmen Sie an einer kurzen Umfrage der Martkforschung teil	Participate in a short market research questionnaire
Intro	Um noch mehr Kunden zu erreichen führen wir gegenwärtig Marktforschungsstudien durch. Wir wären Ihnen sehr dankbar wenn sie sich wenige Minuten Zeit nehmen könnten um uns einige Fragen zu Ihren Fernsehgewohnheiten zu beantworten. Bitte beantworten Sie am Anfang einige Angaben über sich für statistische Zwecke	We want to reach out to even more customers, and are therefore doing some market research. We would appreciate a few minutes of your time to gather your valuable feedback regarding your television watching habits! Please start by sharing with us some details about yourself for statistical purposes

(continued)

Table A.1 (continued)

	Original German wording	English translation
Q1	Wie alt sind Sie?	How old are you?
	– Unter 16 Jahre	– Below 16 years
	– 16–30 Jahre	– 16–30 years
	– 31–45 Jahre	– 31–45 years
	– 45–60 Jahre	– 45–60 years
	– 61–75 Jahre	– 61–75 years
	– 75 Jahre und älter	– 75 years and older
Q2	Sind sie männlich oder weiblich?	Are you male or female?
	– Männlich	– Male
	– Weiblich	– Female
Q3	Wie oft nutzen sie RTL Now?	How often do you use RTL Now?
	– Mehr als einige Male pro Woche	– More than a few times a week
	– Einige Male pro Woche	– A few times a week
	– Ein bis zweimal pro Woche	– Once or twice a week
	– Ein bis zweimal pro Monat	– Once or twice a month
	– Weniger als einmal im Monat	– Less than monthly
	– Ich kenne oder nutze RTL Now nicht	– I do not use or do not know RTL nicht
Q4	Zu welcher Tageszeit schauen sie üblicherweise Fernsehen? Bitte wählen sie alle zutreffenden Anworten	What time of the day do you typically watch TV? Select all that apply
	– Zwischen 6 Uhr und 9 Uhr	– Between 6 a.m. and 9 a.m.
	– Zwischen 9 Uhr und 12 Uhr	– Between 9 a.m. and 12 p.m.
	– Zwischen 12 Uhr und 15 Uhr	– Between 12 p.m. and 3 p.m.
	– Zwischen 15 Uhr und 18 Uhr	– Between 3 p.m. and 6 p.m.
	– Zwischen 18 Uhr und 21 Uhr	– Between 6 p.m. and 9 p.m.
	– Zwischen 21 Uhr und 0 Uhr	– Between 9 p.m. and 12 a.m.
	– Zwischen 0 Uhr und 3 Uhr	– Between 12 a.m. and 3 a.m.
	– Zwischen 3 Uhr und 6 Uhr	– Between 3 a.m. and 6 a.m.
	– Ich schaue kein Fernsehen	– I don't watch TV
Q5	Wie zufrieden sind Sie mit der Programmauswahl bei RTL Television?	How satisfied are you with the program selection at RTL Television?
	– überhaupt nicht zufrieden	– Not at all satisfied
	– Kaum zufrieden	– Slightly satisfied
	– Etwas zufrieden	– Moderately satisfied
	– Sehr zufrieden	– Very satisfied
	– äußerst zufrieden	– Extremely satisfied

(continued)

Table A.1 (continued)

	Original German wording	English translation
Q6	Dürfen wir Sie sofern wir weitere Fragen haben kontaktieren? Falls ja, bitte hinterlassen Sie uns Ihre E-Mailadresse	Can we contact you for further questions? In that case, please share your email address with us
Poll 7		
Title	Zusammenarbeiten für den Klimaschutz	Working together for the climate protection
Subtitle	World Meteorological Organization (WMO)	World Meteorological Organization (WMO)
Intro	Die Weltorganisation für Meteorologie (World Meteorological Organization, WMO) ist eine Sonderorganisation der Vereinten Nationen und spielt eine führende Rolle bei den internationalen Bemühungen zum Klimaschutz. Um zu untersuchen wie effektiv unsere Arbeit ist und an welchen Stellen wir unseren Bekanntheitsgrad verbessern sollten, möchten wir wissen ob und in welchem Ausmaß unsere letztjährigen Kampagnen Sie erreicht haben	World Meteorological Organization (WMO) is a specialized agency of the UN and plays a leading role in international efforts to monitor and protect the environment through its programmes. To investigate how effective our climate work is and where we should improve our visibility, we are interested in hearing how well our campaigns have reached the audience during the year 2014
Q1	Haben Sie während des letzten halben Jahres von WMO Kampagnen gehört oder WMO Kampagnen gesehen?	Have you seen or heard some of WMO's campaigns during the past half a year?
	– Ja	– Yes
	– Nein	– No
	– Ich bin nicht sicher	– I am not sure
Q2	Wie würden Sie Ihre Kenntnisse zum Thema erneuerbare Energien einschätzen?	How would you rate your knowledge on renewable energy?
	– Sagt mir nichts	– It doesn't say anything to me
	– Habe schon davon gehört	– I have heard about it
	– Kenne ich ein wenig	– I am a little bit familiar with the topic
	– Kenne ich gut	– I know the topic moderately well
	– Kenne ich sehr gut	– I know the topic very well
Q3	3. Für wie wichtig halten Sie die Arbeit der WMO in Sachen Klimaschutz?	How important do you find the climate protection work that WMO is conducting?
	– Nicht wichtig	– Not important
	– Weniger wichtig	– Somewhat important
	– Wichtig	– Important
	– Sehr wichtig	– Very important
	– Weiß nicht	– I cannot say
Q4	Was ist Ihr Geschlecht?	Was ist Ihre geschlecht?

(continued)

Table A.1 (continued)

	Original German wording	English translation
Q5	Welche der folgenden Aussagen trifft auf Ihren derzeitigen beruflichen Status bzw. Ihre Ausbildung zu?	Which of the following is true regarding your current employment status and your education?
	– Voll berufstätig	– Fulltime employed
	– Teilweise berufstätig—Teilzeit/stundenweise/zeitweise	– Partial time employed—part time/hourly/temporary
	– Vorübergehend nicht berufstätig, arbeitslos	– Temporarily not employed, unemployed
	– Nicht mehr berufstätig—in Rente/Pension	– No longer working—in retirement/pension
	– Nicht (mehr) berufstätig—Hausfrau/Hausmann	– No (longer) working—housewife/homemaker
	– In Berufsausbildung/Lehre/Wehrpflicht/Zivildienst	– In occupational education/teaching/military/civil service
	– In Schulausbildung—Schüler	– In basic education—pupils and students
	– In Hochschulausbildung—Student	– In higher education—students
	– Nicht berufstätig und nie berufstätig gewesen	– Not employed and never worked
Q6	Können wir Sie für weitere Fragen kontaktieren? Falls ja, bitte teilen Sie uns Ihre E-Mailadresse mit und wir werden sie demnächst kontaktieren?	Can we contact you for further questions? In that case, please share your email address with us and we might contact you in the near future.
Poll 8		
Title	Helfen Sie mit ein besseres Berlin zu ermöglichen!	Help create a Better Berlin!
Subtitle	Berlin betterplace.org	Berlin betterplace.org
Intro	Das Projekt betterplace.org hat bereits einige Kampagnen durchgeführt, um Berlin zu einem besseren Platz zu machen, aber wir sind noch nicht fertig! Um uns beim Helfen zu helfen, möchten wir gerne Ihre Meinung zum Ziel unseres Projektes und zu unserem Bekanntheitsgrad wissen! Am Ende des Fragebogen sind einige Fragen zu Ihrer Person für unsere Statistiken	The project betterplace.org has already had some campaigns to make Berlin a better place, but we are not done yet! To help us help, we are asking you for your opinion on our projects' scope and visibility! At the end of this short questionnaire there will be a few questions about yourself for our statistical information.
Q1	Haben sie bereits von betterplace.org gehört?	Have you heard of betterplace.org before?
	– Nein, habe ich nicht	– No, I have not
	– Ja, online	– Yes, online
	– Ja, ich habe Poster gesehen	– Yes, I have seen posters
	– Ja, ich habe andere Leute darüber reden gehört	– Yes, I have heard people speaking about it
	– Ja, ich habe an andere Stelle davon gehört:	– Yes, I have heard about it in another context:

(continued)

Table A.1   (continued)

	Original German wording	English translation
Q2	An welche betterplace.org Kampagnen erinnern sie sich?	Which betterplace.org campaigns do you remember having seen during the last 12 months?
	– Ich kann mich an keine Kampagne erinnern	I don't remember seeing any campaigns
	– Kampagnen zur Unterstützung der Kältehilfe	Campaigns for supporting the homeless such as Kältehilfe
	– Kampagnen zur Förderung der Barrierefreiheit öffentlicher Gebäude	Campaigns to promote accessibility of public buildings
	– Kampagnen für ein besseres Freizeitangebot an öffentlichen Schulen	Campaigns for improving the freetime activities in public schools
	– Kampagnen zur internationalen Bekämpfung von Ebola	Campaigns for international issues such as Ebola
	– Andere	Other
Q3	Bitte teilen Sie uns mit wie wir uns verbessern können! Jeder Hinweis von Problemen mit dem Bewerben neuer Kampagnen ist willkommen	Leave us a comment on how we could improve! Any comments from issues with advertising to new campaigns are welcome
Q4	Wie alt sind Sie?	Wie alt sind Sie?
Q5	Was ist Ihr höchster Schulabschluss?	Was ist Ihr höchster Schulabschluss?
	– Hauptschulabschluss	– Hauptschulabschluss
	– Mittlere Reife	– Mittlere Reife
	– Abitur (inkl. Fachabitur)	– Abitur (inkl. Fachabitur)
	– Abgeschlossenes Studium (Uni, FH)	– Abgeschlossenes Studium (Uni, FH)
	– kein Schulabschluss	– kein Schulabschluss
Q6	Möglicherweise haben wir weitere Fragen zu Ihrem Feedback. Wenn Sie damit einverstanden sind, bitte teilen sie uns Ihre E-Mailadresse mit.	We might have some further questions regarding your feedback, and would need to contact you for this. If you agree with us contacting you, please share your email address!

The polls had a short title as well as a longer subtitle defining the coarse theme of the poll, and an introduction to the purpose of the poll as well as a possible compensation provided. Then, the six questions included in each poll were presented one after another, with either single choice, multiple choice, or free-text answers options

The mean age of the participants was 28.8 yrs; other demographic details are listed in Table A.3. Occupational information was not collected in this study.

Table A.2 Demographic distributions in Study VII

Demographic		% of participants
Gender	Female	41
	Male	59
Education	Secondary school (Hauptschulabschluss)	0
	Secondary school (Realschulabschluss)	21
	Highschool (Abitur)	36
	University/university of applied sciences	44
Occupation	Student	51
	Employee	33
	Entrepreneur	8
	Unemployed	8

Table A.3 Demographic distributions in Study VIII

Demographic		% of participants
Gender	Female	68
	Male	32
Education	Secondary school (Hauptschulabschluss)	0
	Secondary school (Realschulabschluss)	24
	Highschool (Abitur)	47
	University/university of applied sciences	29

A.5 Study IX: Questionnaire Construction

A survey with pen and paper was conducted to evaluate smartphone users' privacy perceptions ($N = 19$); the study was part of a student project during a Privacy Seminar at the TU Berlin [138]. Convenience sampling was deployed for recruiting the participants with smartphone ownership being the only requirement. The survey was conducted at a location convenient to each participant, such as in a cafe or home environment. Participation was voluntary, and participants received no incentives. The scales deployed were PUSU, UNAUT, ACCU, COLL, RISK, BENE, NORM, and DFP (WILL).

The participants mean age was 34.2 yrs; the other demographic details are listed in Table A.4.

A.6 Study X: Questionnaire Construction

This survey was conducted as an online survey using the Limesurvey platform. The survey was part of a study investigating the most commonly used communication means among online users. The participants ($N = 72$) were recruited through the

Table A.4 Demographic distributions in Study IX

Demographic		% of participants
Gender	Female	57
	Male	42
Education	Secondary school (Hauptschulabschluss)	1
	Secondary school (Realschulabschluss)	21
	Highschool (Abitur)	37
	University/university of applied sciences	37
Occupation	Student	5
	Employee	79
	Entrepreneur	16
	Unemployed	0

Table A.5 Demographic distributions in Study VIII

Demographic		% of participants
Gender	Female	64
	Male	36
Education	Secondary school (Hauptschulabschluss)	0.5
	Secondary school (Realschulabschluss)	11
	Highschool (Abitur)	41
	University/university of applied sciences	44

Prometei portal, and compensated with a 5€ voucher for the participation. The scales deployed in this study were RISK, BENE, and NORM.

The mean age of participants was 30.8 yrs ($SD = 9.47$); the other demographics are detailed in Table A.5.

A.7 Study XI: Questionnaire Construction

This survey was conducted as part of a study investigating smartphone users' awareness of information privacy practices on smartphone applications ($N = 68$) [138]. The participants for this study were recruited through online SNSs. The survey was conducted using the Limesurvey platform. No incentive was given for participation in this study. The scales deployed in the study were RISK, BENE, and DFP (WILL). The demographic distributions are listed in Table A.6.

Table A.6 Demographic distributions in Study XI

Demographic		% of participants
Gender	Female	41
	Male	59
Age group	18–22	34
	23–27	40
	28–32	9
	33–37	6
	38–42	4
	43–47	34
	48–52	1
	53–57	3
Education	Secondary school (Realschulabschluss)	6
	Highschool (Abitur)	68
	University/university of applied sciences	26
Occupation	Student	57
	Employee	37
	Entrepreneur	3
	Unemployed	3

A.8 Study XII: Questionnaire Construction

A survey was conducted in September 2016 in order to develop the LBS Knowledge scale. The study ($N = 30$) was conducted using the Limesurvey platform, and participants were given a 4€ Amazon voucher as an incentive for participation. The questions included in this study are listed in Table A.7

The mean age of the participants was 31.5 yrs ($SD = 7.43$); the other demographics are detailed in Table A.8.

A.9 Full List of Initial Items

The following scales have been translated from German to English for reporting, except the privacy concern scale, which was originally reported in English. For each scale, the question order was randomized for each participant. The items that were reverse-coded prior to analysis in order to match with the scale direction are marked with an asterisk (*) (Table A.9).

Table A.7 List of initial questions included in the LBS Knowledge scale

Code	Question	Answer options
LBS1	Um die Standortdienste auf meinem Smartphone auszuschalten, …	… sollte ich in die Einstellungen gehen und die Standortdienste deaktivieren
		… sollte ich das GPS ausschalten
		… sollte ich das WLAN ausschalten
		… sollte ich mein Handy auf Flugmodus stellen
LBS2	Wenn ich das GPS ausschalte …	… kann mein Handy sich nicht mehr mit den GPS Satelliten verbinden und die Ortung wird ungenauer
		… kann ich mich mit meinem Handy nicht mehr lokalisieren. kann ich mich mit meinem Handy nicht mehr lokalisieren
		… werden keine Ortsdaten mehr an Handy oder Apphersteller gesendet
		… verbindet sich mein Handy nur noch sporadisch mit GPS Satelliten
LBS3	Was steht in Datenschutzrichtlinien?	Ob und wie Firmen persönliche Daten verarbeiten
		Was der Nutzer tun muss, um seine Daten zu schützen
		Wie persönliche Daten im Allgemeinen eingeordnet werden
		Das persönliche Daten immer anonymisiert verarbeitet werden
LBS4	Dürfen ortsbasierte Daten mit anderen Daten kombiniert und an Dritte weitergegeben werden?	Ja, aber nur wenn ich dem einmalig ausdrücklich zugestimmt habe
		Ja
		Nein
		Ja, aber ich muss der Weitergabe jedes Mal ausdrücklich zustimmen
LBS5	Für eine rechtmäßige Einwilligung der Verarbeitung von Daten muss sichergestellt sein, dass …	… Alle Antworten sind richtig
		… sie jederzeit widerrufbar ist, worauf ich hingewiesen werden muss
		… sie von mir bewusst und eindeutig erteilt wird
		… der Inhalt jederzeit für mich verfügbar ist
LBS6	Betriebssystemherstellern (z.B. Android mit Google oder Apple mit iOS) ist es nicht erlaubt meine Standortdaten zu sammeln …	… für Geheimdienste
		… um häufig besuchte Orte herauszufinden
		… um mir Vorschläge für meine Umgebung zu machen
		… für Verkehrsinformationen

(continued)

Table A.7 (continued)

Code	Question	Answer options
LBS7	Welche der folgenden Antworten ist korrekt?	Apps können meinen Standort verwenden, obwohl sie gar nicht geöffnet sind.
		Apps können nur auf meinen Standort zugreifen, wenn sie geöffnet sind
		Wenn mein Handy ausgeschaltet ist, können Google (oder Apple) trotzdem auf meinen Standort zugreifen
		App Hersteller dürfen meine Standortdaten benutzen, ohne mich darauf hinzuweisen, dass Sie den Standort verwenden
LBS8	Wenn ich die Standortdienste deaktiviert habe, …	…dann hat nur mein Mobilfunkanbieter trotzdem Zugriff auf meinen Standort
		…dann hat nur mein Betriebssystemhersteller (Apple/Google) trotzdem Zugriff auf meinen Standort.
		…dann haben Betriebssystemhersteller und Apps trotzdem Zugriff auf meinen Standort
		…Keine der Antworten ist korrekt
AC1	Wenn Sie diese Frage lesen wählen Sie Bitte die Antwort, die mit "Ortsbasierte Dienste" beginnt. …	…Ortsbasierte Dienste helfen mir mich zu orientieren
		…Ich nutze ortsbasierte Dienste gerne um mich zurecht zu finden
		…Keine der Antworten
		…Wenn sie Notfallrufe absenden und wenn Sie ihr Telefon als verloren melden
Google	The following questions are shown only to the participants using phones with the Android operating system.	
LBS9	Betriebssystemherstellern (z.B. Google mit Android) ist es nicht erlaubt meine Standortdaten zu sammeln …	…für Geheimdienste.
		…um häufig besuchte Orte herauszufinden
		…um mir Vorschläge für meine Umgebung zu machen
		…für Verkehrsinformationen
LBS10	Betriebssystemhersteller (z.B. Google mit Android) sammeln meine Standortdaten für …	…Alle Antworten sind richtig
		…Werbung
		…Analysezwecke
		…Verkehrsinformationen
LBS11	Wenn die Standortdienste aktiviert sind, wann kann Google Zugriff auf meinen Standort haben?	Die ganze Zeit
		Nur wenn ich den Standort ändere
		Nur wenn ich eine App öffne welche den Standort benötigt (z.B. Maps)
		Wenn ich GPS eingeschaltet habe
LBS12	Wann verwendet Google meinen Standort, obwohl ich die Standortdienste deaktiviert habe?	Gar nicht
		Wenn ich Notfallrufe absende und wenn ich mein Telefon als verloren melde
		Nur wenn ich mein Telefon als verloren melde
		Jederzeit

(continued)

Table A.7 (continued)

Code	Question	Answer options
LBS13	Wenn ich die Standortdienste aktiviert habe werden diese Daten von Google wie gespeichert?	Google speichert die Daten verschlüsselt und mit meiner Person verknüpft auf ihren Servern
		Google speichert keine Standortdaten auf ihren Servern, sondern nur auf meinem Smartphone
		Google speichert die Standortdaten unverschlüsselt, aber anonymisiert auf ihren Servern
		Keine der Antworten ist korrekt
LBS14	Ortsbasierte Daten können mit Daten verknüpft werden, welche mir eindeutig zuzuordnen sind (es kann also genau gesagt werden, wo ich war). Dies gilt für folgende Unternehmen:	Google und Appanbieter
		Ortsbasierte Daten dürfen nicht mit Daten gespeichert werden, welche einer Person eindeutig zugeordnet werden können
		Google
		Appanbieter
AC2	Wenn Sie diese Frage lesen, wählen Sie bitte die Antwort in der "Apple" vorkommt.	Google nutzt meine Standortdaten um Apple Kunden wegzunehmen
		Google nutzt meine Standortdaten für Werbezwecke
		Google nutzt meine Standortdaten um herauszufinden wo ich wohne
		Keine der Antworten ist korrekt
Apple	The following questions are shown only to the participants using phones with the iOS operating system.	
LBS15	Betriebssystemherstellern (z.B. Apple mit iOS) ist es nicht erlaubt meine Standortdaten zu sammeln …	… für Geheimdienste
		… um häufig besuchte Orte herauszufinden
		… um mir Vorschläge für meine Umgebung zu machen
		… für Verkehrsinformationen
LBS16	Betriebssystemhersteller (z.B. Apple mit iOS) sammeln meine Standortdaten für …	… Alle Antworten sind richtig
		… Werbung
		… Analysezwecke
		… Verkehrsinformationen
LBS17	Wenn die Standortdienste aktiviert sind, wann kann Apple Zugriff auf meinen Standort haben?	Die ganze Zeit
		Nur wenn ich den Standort ändere
		Nur wenn ich eine App öffne welche den Standort benötigt (z.B. Maps)
		Wenn ich GPS eingeschaltet habe
LBS18	Wann verwendet Apple meinen Standort, obwohl ich die Standortdienste deaktiviert habe?	Wenn ich Notfallrufe absende und wenn ich mein Telefon als verloren melde
		Gar nicht
		Nur wenn ich mein Telefon als verloren melde
		Jederzeit

(continued)

Table A.7 (continued)

Code	Question	Answer options
LBS19	Wenn ich Standortdienste aktiviert habe werden diese Daten von Apple wie gespeichert?	Apple speichert Standortdaten verschlüsselt und anonymisiert auf ihren Servern und auf meinem Smartphone
		Apple speichert Standortdaten unverschlüsselt, aber anonymisiert auf ihren Servern
		Apple speichert die Daten mit meiner Person verknüpft auf ihren Servern
		Keine der Antworten ist korrekt
LBS20	Ortsbasierte Daten können mit Daten verknüpft werden, welche mir eindeutig zuzuordnen sind (es kann also genau gesagt werden, wo ich war). Dies gilt für folgende Unternehmen:	Appanbieter
		Apple und Appanbieter
		Ortsbasierte Daten dürfen nicht mit Daten gespeichert werden, welche einer Person eindeutig zuzuordnen sind
		Apple
AC3	Wenn Sie diese Frage lesen, wählen Sie bitte die Antwort in der "Google" vorkommt.	Apple nutzt meine Standortdaten um Google Kunden wegzunehmen
		Apple nutzt meine Standortdaten für Werbezwecke
		Apple nutzt meine Standortdaten um herauszufinden wo ich wohne
		Keine der Antworten ist korrekt

The items in **boldface** are included in the final questionnaire. The question order, as well as the answer order, was randomized. For each question, the first answer option is correct. The items AC1, AC2, and AC3 are trapping questions used for attention check

Table A.8 Demographic distributions in Study XII

Demographic		% of participants
Gender	Female	63
	Male	37
Education	Secondary school (Hauptschulabschluss)	5
	Secondary school (Realschulabschluss)	11
	Highschool (Abitur)	47
	University/University of applied sciences	37
Occupation	Student	53
	Employee	42
	Entrepreneur	0
	Unemployed	11

Table A.9 Full list of items used in the development of the questionnaire for Location Privacy Beliefs and Attitudes (LPBA). The final questionnaire items are in bold-face. The reverse-coded items are marked with an asterisk (*). A translation from the original German text into English is provided

Abbreviation	Original German wording	English translation
PUSU1	Wenn Leute einer mobilen App aus einem bestimmten Grund persönliche Informationen geben, sollten die Firmen, die die App betreiben, die Information niemals zu irgendeinem anderen Zweck nutzen	When people give personal information to a mobile application for some reason, the company behind the application should never use the information for any other reason
PUSU2	Firmen, die mobile Apps betreiben, sollten persönliche Informationen auf keinen Fall nutzen, es sei denn dies wurde von den Personen, die die Information bereitgestellt haben, genehmigt	Companies behind mobile applications should not use personal information for any purpose unless it has been authorized by the individuals who provided information
PUSU3	Mobile Apps sollten den Standort eines Nutzers nur verfolgen, wenn es für ihre Funktionalität erforderlich ist	Mobile applications should track users' location only if it is required for their functionality
PUSU4	Firmen, die mobile Apps betreiben, sollten niemals die persönlichen Informationen aus Ihren Datenbanken an andere Firmen verkaufen	Companies behind mobile applications should never sell the personal information in their databases to other companies
PUSU5	Mobile Apps sollten nur nach denjenigen Berechtigungen fragen, die für ihre Funktionalität erforderlich sind	Mobile applications should only ask for the permissions that are required for their functionality
PUSU6	Firmen, die mobile Apps betreiben, sollten niemals persönliche Informationen mit anderen Firmen teilen, es sei denn es wurde von der Person, die die Information bereitgestellt hat, genehmigt	Companies behind mobile applications should never share personal information with other companies unless it has been authorized by the individuals who provided the information
PUSU7	Datenbanken von mobilen Apps, die persönliche Informationen beinhalten sollten vor unbefugtem Zugang geschützt werden—egal wieviel es kostet	Mobile applications' databases that contain personal information should be protected from unauthorized access—no matter how much it costs
PUSU8	Firmen, die mobile Apps betreiben, sollten mehr Zeit und Mühe darauf verwenden, unbefugtem Zugang zu persönlichen Informationen zu vermeiden	Companies behind mobile applications should devote more time and effort to preventing unauthorized access to personal information
PUSU9	Mobile Apps sollten keine unabhängigen Entscheidungen über meine persönlichen Informationen ohne meine Zustimmung treffen	Mobile applications should not make independent decisions without my consent about my personal information
PUSU10	Es stört mich so vielen mobile Apps meine persönlichen Informationen zu geben	It bothers me to give personal information to so many mobile applications
PUSU11	Ich glaube, dass Firmen, die Apps für Mobilgeräte betreiben, daran interessiert sind meine Standortdaten zu Marketingzwecken zu verkaufen	I believe that companies behind mobile applications are interested in selling my location data for marketing purposes

(continued)

Table A.9 (continued)

Abbreviation	Original German wording	English translation
PUSU12	Firmen, die Apps für Mobilgeräte betreiben, sollten nur diejenigen Nutzerdaten sammeln, die für die Funktionalität der App erforderlich sind	Companies behind mobile applications should collect only the user data that is required for the functionality of the application
PUSU13	Firmen, die mobile Apps betreiben, sollten niemals die Standortdaten Ihrer Nutzer zu Marketingzwecken verkaufen.	Companies behind mobile applications should not sell the users' location data for marketing purposes.
UNAUT1*	Ich glaube, dass keine Risiken damit verbunden sind, wenn mobile Apps Standortinformationen sammeln, die anonym sind	I believe that there are no risks involved when mobile applications collect location information that is anonymous.
UNAUT2*	Ich glaube, dass mobile Apps nur den Standort der Nutzer verfolgen, wenn es für ihre Funktionalität erforderlich ist	I believe that mobile applications track users' location only if it is required for their functionality.
UNAUT3*	Ich glaube, dass mobile Apps nur diejenigen Nutzerdaten sammeln, die für die Funktionalität der App benötigt werden	I believe that mobile applications collect only the user data that is required for the functionality of the application.
UNAUT4*	Ich glaube, dass mobile Apps nur nach denjenigen Berechtigungen fragen, die für ihre Funktionalität erforderlich sind	I believe that mobile applications only ask for the permissions that are required for their functionality.
UNAUT5	Firmen, die mobile Apps betreiben, sollten mehr Maßnahmen ergreifen um sicherzustellen, dass Unbefugte keinen Zugang zu persönlichen Informationen auf ihren Computern haben.	Companies behind mobile applications should take more steps to make sure that unauthorized people cannot access personal information on their computers.
UNAUT6	Firmen, die mobile Apps betreiben, sollten mehr Maßnahmen ergreifen um sicherzustellen, dass persönliche Informationen in ihren Karteien richtig sind.	Companies behind mobile applications should take more steps to make sure that the personal information in their files is accurate.
ACCU1	Alle persönlichen Informationen, die in den Datenbanken von mobile Apps gespeichert sind, sollten genau auf ihre Richtigkeit hin überprüft werden—egal wieviel es kostet	All the personal information in mobile application databases should be double-checked for accuracy—no matter how much it costs.
ACCU2	Firmen, die mobile Apps betreiben, sollten bessere Verfahren haben um Fehler bei persönlichen Informationen zu korrigieren	Companies behind mobile applications should have better procedures to correct errors in personal information.
ACCU3	Firmen, die mobile Apps betreiben, sollten mehr Zeit und Mühe darauf verwenden die Richtigkeit von persönlichen Informationen in ihren Datenbanken zu überprüfen	Companies behind mobile applications should devote more time and effort to verifying the accuracy of the personal information in their databases.
ACCU4	Ich mache mir Sorgen, dass mobile Apps unabhängige Entscheidungen über meine persönlichen Informationen ohne meine vorherige Zustimmung treffen.	I am worried that mobile applications make independent decisions without my consent about my personal information.

(continued)

Table A.9   (continued)

Abbreviation	Original German wording	English translation
COLL1	Ich mache mir sorgen, dass mobile Apps zu viele persönliche Informationen über mich sammeln	I'm concerned that mobile applications are collecting too much personal information about me
COLL2	Für gewöhnlich stört es mich, wenn mobile Apps nach meinen persönlichen Informationen fragen	It usually bothers me when mobile applications ask me for personal information
COLL3	Wenn mich mobile Apps nach meinen persönlichen Informationen fragen, denke ich manchmal zweimal darüber nach bevor ich sie herausgebe	When mobile applications ask me for personal information, I sometimes think twice before providing it
RISK1	Die Nutzung ortsbasierter Services ist riskant	Using location-based applications is risky
RISK2	Die Nutzung ortsbasierter Services beinhaltet das Risiko gestalkt zu werden	Using location-based services involves the risk of getting stalked
RISK3	Ich bin besorgt, dass die Nutzung ortsbasierter Services dazu führt meine Hausadresse zu verraten	I am worried that using location-based services would lead to my home location being revealed
RISK4	Ich bin besorgt, dass ich von meinem Chef geortet werden könnte, wenn ich Location-based Services nutze	I am worried that if I use location-based services, I might get tracked by my boss
RISK5	Ich bin besorgt, dass ich vom Staat geortet werden könnte, wenn ich Location-based Services nutze	I am worried that if I use location-based services, I might get tracked by the government
RISK6	Ich bin besorgt, dass die Nutzung ortsbasierter Services zu unaufgeforderten Marketing führt	I am worried that using location-based services would lead to unsolicited marketing
RISK7	Ich bin besorgt, dass die Nutzung ortsbasierter Services das Risiko birgt, Opfer von Identitätsdiebstahl zu werden	I am worried that using location-based services involves the risk of becoming a victim of identity theft
RISK8	Ich bin besorgt, dass Fremde zu viel über meine Aktivitäten wissen könnten, wenn ich ortsbasierte Services nutze	I am worried that if I use location-based services, strangers might know too much about my activities
RISK9	Die Nutzung ortsbasierter Services stellt eine Gefahr für meine persönliche Sicherheit dar	Using location-based services poses a threat to my personal safety
RISK10*	Ich glaube, dass keine Risiken damit verbunden sind, wenn mobile Apps Standortinformationen sammeln, die anonym sind	I believe that there are no risks involved when mobile applications collect location information that is anonymous
RISK11	Ich glaube, dass Firmen, die Apps für Mobilgeräte betreiben, daran interessiert sind meine Standortdaten zu Marketingzwecken zu verkaufen	I believe that companies behind mobile applications are interested in selling my location data for marketing purposes
RISK12*	Ich glaube, dass es in meinem Bekanntenkreis keine Opfer von Telefonüberwachung gibt	I believe that within my circles, there are no victims of phone surveillance

(continued)

Table A.9 (continued)

Abbreviation	Original German wording	English translation
RISK13*	Ich glaube, dass mobile Apps nur den Standort der Nutzer verfolgen, wenn es für ihre Funktionalität erforderlich ist	I believe that mobile applications track users' location only if it is required for their functionality
RISK14	Ich glaube, wenn ich meine Standortdaten an mobile Apps weitergebe, bin ich einem Stalking-Risiko ausgesetzt	Giving out my location information to mobile applications involves the risk of stalking
RISK15	Ich lese mir alle Berechtigungen, die von einer App gefordert werden, sorgfältig durch, bevor ich die App installiere	I read carefully all asked permissions before installing a mobile application
BENE1	Die Nutzung ortsbasierter Services macht Spaß	Using location-based services is fun
BENE2	Die Nutzung ortsbasierter Services ist praktisch	Using location-based services is practical
BENE3	Die Nutzung ortsbasierter Services ist nützlich	Using location-based applications is useful
BENE4	Die Nutzung ortsbasierter Services ermöglicht es mir Aufgaben schneller durchzuführen	Using location-based services enables me to perform tasks faster
BENE5	Die Nutzung ortsbasierter Services macht die Kommunikation schneller	Using location-based services makes communication faster
BENE6	Die Nutzung orstabsierter Services macht die Kommunikation einfacher	Using location-based services simplifies communication
BENE7	Ortsbasierte Services verbessern mein soziales Leben	Location-based services enhance my social life
BENE8	Standortbezogene Dienste erzeugen eine sichere Gesellschaft	Location-based services create a secure society
BENE9	Das Benutzen standortbezogener Dienste vereinfacht das Ausmachen von Meetings	Using location-based services makes it easy to schedule meetings
BENE10	Standortbezogene Dienste verbessern mein Sozialleben	Location-based services improve my social life
BENE11	Standortbezogene Dienste sind vorteilhaft für personalisierte Suchen	Location-based services are beneficial for personalized searches
BENE12	Standortbezogene Dienste sind vorteilhaft für verhaltensbasierte Werbung	Location-based services are beneficial for behaviour-based advertising
BENE13	Die Nutzung ortsbasierter Services macht die Kommunikation einfacher	The use of location-based services makes communication easier
BENE14	Das Benutzen standortbezogener Dienste verbessert den Eindruck den andere von mir haben	Using location-based services improves the impression others have of me
NORM1	Es ist mir wichtig, was Menschen, die mir etwas bedeuten, von mir halten	It is important to me what people who I care about think of me
NORM2	Leute in meinem sozialen Umfeld erwarten, dass ich ortsbasierte Apps nutze	People in my social circles expect me to use location-based applications
NORM3	Leute in meinem sozialen Umfeld nutzen Location-based Services	People in my social circles use location-based applications

(continued)

Table A.9 (continued)

Abbreviation	Original German wording	English translation
NORM4	Ich glaube, dass Leute in meinem sozialen Umfeld denken, die Nutzung Location-based Services Risiken birgt	I believe that people in my social circles think that using location-based applications is risky
NORM5	Jeder nutzt ortsbasierte Apps	Everybody uses location-based applications
NORM6	Menschen, die Location-based Services nutzen, werden als fortschrittlich betrachtet	People who use location-based applications are considered progressive
NORM7	Menschen, die mir etwas bedeuten und denen ich etwas bedeute, denken dass ich ortsbasierte Apps nutzen sollte	People who I care about and who care about me think that I should use location-based applications
NORM8	Menschen, die mir wichtig sind, denken ich sollte ortsbasierte Apps nutzen	People who are important to me think that I should use location-based applications
NORM9	Menschen, die mir etwas bedeuten und denen ich etwas bedeute, denken dass es bestimmte Vorteile hat ortsbasierte Apps zu nutzen	People who I care about and who care about me think that there are certain benefits in using location-based applications
NORM10	Menschen, die mir wichtig sind, denken ich sollte Location-based Services nutzen	People who are important to me think that there are certain benefits in using location-based applications
NORM11	Menschen, die mir etwas bedeuten und denen ich etwas bedeute, denken dass es bestimmte Risiken in sich birgt Location-based Services zu nutzen	People who I care about and who care about me think that there are certain risks involved in using location-based applications
NORM12	Menschen, die mir wichtig sind, denken dass es bestimmte Risiken in sich birgt, Location-based Services zu nutzen	People who are important to me think that there are certain risks involved in using location-based applications
DFP1 = WILL1*	Am wichtigsten ist es für mich, meine Privatsphäre zu schützen	It is the most important thing for me to protect my privacy
DFP2* = WILL2	Ich fühle mich wohl anderen Leuten, einschließlich Fremden, persönliche Informationen über mich selbst zu geben	I'm comfortable telling other people, including strangers, personal information about myself
DFP3 = WILL3*	Ich versuche die Häufigkeit, personenbezogene Angaben über mich anzugeben, zu minimieren	I try to minimize the number of times I have to provide personal information about myself
DFP4* = WILL4	Ich fühle mich wohl, mit anderen Leuten Informationen über mich selbst zu teilen, sofern sie mir keinen Grund geben es nicht zu tun	I am comfortable sharing information about myself with other people unless they give me reason not to
DFP5* = WILL5	Ich habe nichts zu verbergen und habe deshalb kein Problem damit, wenn andere Personen persönliche Dinge über mich wissen	I have nothing to hide, so I am comfortable with people knowing personal information about me
DFP6 = WILL6*	Ich versuche das Thema zu wechseln, wenn die Leute zu viel über mich fragen	I try to change the topic of a conversation if people start asking too much about me

A.10   Trust

All items for measuring trust within the Study IV are listed in Table A.10. Also a translation from the original German text into English is provided.

Table A.10 List of questions included in the Trust scales. The question order was randomized. The answers were given on a seven-point end-labelled Likert scale

Code	Original German item	English translation
Trustworthiness of Crowdee		
TRUST1	Für wie vertrauenswürdig halten Sie Crowdee?	How trustworthy do you find Crowdee?
TRUST2	Für wie glaubwürdig halten Sie Crowdee?	How reliable do you find Crowdee?
TRUST3	Im Allgemeinen wie riskant empfinden Sie es Standortdaten an Crowdee weiterzugeben?	In general, how risky do you find it to give location information to Crowdee?
TRUST4	Wie besorgt sind sie dass Crowdee Ihnen Schaden zufügen könnte wenn sie im Besitz Ihrer Standortdaten sind?	How concerned are you that Crowdee could harm you if it had your location data?
Trustworthiness of a trusted advertiser		
TRUST5	Für wie vertrauenswürdig halten Sie das Deutsche Rote Kreuz?	How trustworthy do you find the German Red Cross?
TRUST6	Für wie glaubwürdig halten Sie das Deutsche Rote Kreuz?	How reliable do you find the German Red Cross?
TRUST7	Im Allgemeinen wie riskant empfinden Sie es Standortdaten an das Deutsche Rote Kreuz weiterzugeben?	In general, how risky do you find it to give location information to the German Red Cross?
TRUST8	Wie besorgt sind sie dass das Deutsche Rote Kreuz Ihnen Schaden zufügen könnte wenn sie im Besitz Ihrer Standortdaten sind?	How concerned are you that the German Red Cross could harm you if it had your location data?
Trustworthiness of an untrusted advertiser		
TRUST9	Für wie vertrauenswürdig halten Sie die BILD Zeitung	How trustworthy do you find the Bild Magazine?
TRUST10	Für wie glaubwürdig halten Sie die BILD Zeitung?	How reliable do you find the Bild Magazine?
TRUST11	Im Allgemeinen wie riskant empfinden Sie es Standortdaten an die BILD Zeitung weiterzugeben?	In general, how risky do you find it to give location information to the Bild Magazin?
TRUST12	Wie besorgt sind sie dass die BILD Zeitung Ihnen Schaden zufügen könnte wenn sie im Besitz Ihrer Standortdaten sind?	How concerned are you that the Bild Magazine could harm you if it had your location data?
Trustworthiness of other Crowdee users		
TRUST13	Für wie vertrauenswürdig halten Sie die anderen Crowdee-Nutzer?	How trustworthy do you find the other Crowdee users?

(continued)

Table A.10   (continued)

Code	Original German item	English translation
TRUST14	Für wie glaubwürdig halten Sie die anderen Crowdee-Nutzer?	How reliable do you find the other Crowdee users?
TRUST15	Im Allgemeinen wie riskant empfinden Sie es Standortdaten an andere Crowdee-Nutzer weiterzugeben?	In general, how risky do you find it to give location information to other Crowdee users?
TRUST16	Wie besorgt sind sie dass Crowdee-Nutzer Ihnen Schaden zufügen könnten wenn sie im Besitz Ihrer Standortdaten sind?	How concerned are you that Crowdee users could harm you if it had your location data?

The German end labels for items TRUST1, TRUST5, TRUST9, and TRUST13 were "Sehr vertrauenswürdig" and "Überhaupt nicht vertrauenswürdig"; for items TRUST2, TRUST6, TRUST10, and TRUST14 "Sehr glaubwürdig" and "Überhaupt nicht glaubwürdig"; for items TRUST3, TRUST7, TRUST11, and TRUST15 "Sehr riskant" and "Überhaupt nicht riskant"; and for items TRUST4, TRUST8, TRUST12, and TRUST16 "Sehr besorgt" and "Überhaupt nicht besorgt"

References

1. Ackerman, M.S., Cranor, L.F., Park, F., Reagle, J.: Privacy in E-commerce: examining user scenarios and privacy preferences. In: Proceedings of the 1999 ACM Conference on Electronic Commerce, 1998, pp. 1–8. ACM, New York (1999)
2. Acquisti, A.: Privacy in electronic commerce and the economics of immediate gratification. In: 5th ACM conference on Electronic Commerce, p. 21. ACM Press, New York (2004). http://portal.acm.org/citation.cfm?doid=988772.988777%5Cnhttp://dl.acm.org/citation.cfm?id=988777
3. Acquisti, A.: Nudging privacy: the behavioral economics of personal information. In: Digital Enlightenment Yearbook 2012, pp. 193–197. IOS Press, Amsterdam (2012)
4. Acquisti, A., Brandimarte, L., Loewenstein, G.: Privacy and human behavior in the age of information. Science **347**(6221), 509–14 (2015). http://science.sciencemag.org/content/347/6221/509.abstract
5. Acquisti, A., Friedman, A., Telang, R.: Is there a cost to privacy breaches? An event study—viewcontent.cgi. In: International Conference on Information Systems (ICIS) (2006). https://doi.org/10.1.1.73.2942
6. Acquisti, A., Grossklags, J.: Privacy and rationality in individual decision making. IEEE Secur. Priv. **3**(1) , 26–33 (2005)
7. Acquisti, A., John, L.K., Loewenstein, G.: What is privacy worth? J. Legal Stud. **42**(2), 249–274 (2013)
8. Agarwal, R., Prasad, J.: A conceptual and operational definition of personal innovativeness in the domain of information technology. Inf. Syst. Res. **9**(2), 101–215 (1998). https://doi.org/10.1287/isre.9.2.204
9. Ajzen, I.: The theory of planned behavior. Organ. Behav. Hum. Decis. Process. **50**(2), 179–211 (1991). https://doi.org/10.1016/0749-5978(91)90020-T
10. Almuhimedi, H., Schaub, F., Sadeh, N., Adjerid, I., Acquisti, A., Gluck, J., Cranor, L.F., Agarwal, Y.: Your location has been shared 5398 times!: a field study on mobile app privacy nudging. In: Proceedings of the 33rd Annual ACM Conference on Human Factors in Computing Systems, pp. 787–796. ACM, New York (2015). https://doi.org/10.1145/2702123.2702210
11. Altman, I.: Privacy regulation: culturally universal or culturally specific? J. Soc. Issues **33**(3), 66–84 (1977). https://doi.org/10.1111/j.1540-4560.1977.tb01883.x
12. Altman, I., Taylor, D.A.: Social Penetration: the Development of Interpersonal Relationships, vol. 75. Holt, Rinehart & Winston, Oxford (1973)

13. Andrés, M.E., Bordenabe, N.E., Chatzikokolakis, K., Palamidessi, C.: Geo-Indistinguishability: differential privacy for location-based systems. In: CCS'13, vol. abs/1212.1 (2013)

14. Attal, F., Mohammed, S., Dedabrishvili, M., Chamroukhi, F., Oukhellou, L., Amirat, Y.: Physical human activity recognition using wearable sensors. Sensors 15(12), 31314–31338 (2015). https://doi.org/10.3390/s151229858

15. AV Test GmbH The Independent IT-Security Institute: Security Report 2016/2017 (2016). https://www.av-test.org/fileadmin/pdf/security_report/AV-TEST_Security_Report_2016-2017.pdf

16. Balebako, R., Jung, J., Lu, W., Cranor, L.F., Nguyen, C.: "Little brothers watching you": raising awareness of data leaks on smartphones. In: Proceedings of the Ninth Symposium on Usable Privacy and Security (SOUPS '13), pp. 12:1–12:11 (2013). https://doi.org/10.1145/2501604.2501616

17. Balebako, R., Leon, P.G., Almuhimedi, H., Kelley, P.G., Mugan, J., Acquisti, A., Cranor, L.F., Sadeh, N.: Nudging users towards privacy on mobile devices. In: CEUR Workshop Proceedings, vol. 722, pp. 23–26. CEUR-WS (2011)

18. Bansal, G., Zahedi, F.M., Gefen, D.: The impact of personal dispositions on information sensitivity, privacy concern and trust in disclosing health information online. Decis. Support. Syst. 49(2), 138–150 (2010)

19. Barak, O., Cohen, G., Gazit, a., Toch, E.: The price is right? Economic value of location sharing. In: 2nd ACM International Workshop on Mobile Systems for Computational Social Science, pp. 891–899 (2013). https://doi.org/10.1145/2494091.2497343

20. Barkhuus, L.: Privacy in location-based services , concern vs. coolness. In: Proceedings of MobileHCI 2004 (2004)

21. Barkhuus, L., Brown, B., Bell, M., Hall, M., Sherwood, S., Chalmers, M.: From awareness to repartee: Sharing Location within Social Groups. In: Proceeding of the Twenty-Sixth Annual CHI Conference on Human Factors in Computing Systems - CHI '08, p. 497. ACM Press, New York (2008). https://doi.org/10.1145/1357054.1357134

22. Baron, R.M., Kenny, D.A.: The moderator-mediator variable distinction in social the moderator-mediator variable distinction in social psychological research: conceptual, strategic, and statistical considerations. J. Pers. Soc. Psychol. 51(6), 1173–1182 (1986). https://doi.org/10.1037/0022-3514.51.6.1173

23. Bélanger, F., Carter, L.: Trust and risk in e-government adoption. J. Strateg. Inf. Syst. 17(2), 165–176 (2008)

24. Belanger, F., Hiller, J.S., Smith, W.J.: Trustworthiness in electronic commerce: the role of privacy, security, and site attributes. J. Strateg. Inf. Sys. 11(3–4), 245-270 (2002). https://doi.org/10.1016/S0963-8687(02)00018-5

25. Bellman, S., Johnson, E.J., Kobrin, S.J., Lohse, G.L.: International differences in information privacy concerns: a global survey of consumers. Inf. Soc. Int. J 20(2015), 313–324 (2004). https://doi.org/10.1080/01972240490507956

26. Bentham, J.: The panopticon writings. In: The Panopticon Writings (1995). https://doi.org/10.1007/s00287-006-0116-6

27. Bentler, P.M., Bonnett, D.G.: Significance tests and goodness of fit in the analysis of covariance structures. Psychol. Bull. 88(3), 588–606 (1980). https://doi.org/10.1037/0033-2909.88.3.588

28. Beresford, A.R., Stajano, F.: Location privacy in pervasive computing. IEEE Pervasive Comput. 2(1), 46–55 (2003). https://doi.org/10.1109/MPRV.2003.1186725

29. Bordenabe, N.E., Chatzikokolakis, K., Palamidessi, C.: Optimal geo-indistinguishable mechanisms for location privacy. In: Proceedings of the 2014 ACM SIGSAC Conference on Computer and Communications Security - CCS '14 (2014). https://doi.org/10.1145/2660267.2660345

30. Bortz, J., Döring, N.: Forschungsmethoden und Evaluation für Human- und Sozialwissenschaftler (2006). https://doi.org/10.1007/978-3-540-33306-7

31. Brown, B., Taylor, A.S., Izadi, S., Sellen, A., Kaye, J., Eardley, R.: Locating family values: a field trial of the whereabouts clock. In: Proceedings of the 9th International Conference on Ubiquitous Computing (UbiComp'07) (2007). https://doi.org/10.1007/978-3-540-74853-3

32. Brush, A.J., Krumm, J., Scott, J.: Exploring end user preferences for location obfuscation, location-based services, and the value of location. Proceedings of the 12th ACM international conference on Ubiquitous computing - Ubicomp '10, p. 95. ACM, New York (2010). https://doi.org/10.1145/1864349.1864381

33. Byrne, B.M.: Structural Equation Modeling with AMOS (2010). https://doi.org/10.4324/9781410600219

34. Chin, E., Felt, A.P., Sekar, V., Wagner, D.: Measuring user confidence in smartphone security and privacy. In: Proceedings of the Eighth Symposium on Usable Privacy and Security - SOUPS '12, vol. 1, p. 1 (2012). http://dl.acm.org/citation.cfm?doid=2335356.2335358

35. Consolvo, S., Smith, I.E., Matthews, T., LaMarca, A., Tabert, J., Powledge, P.: Location disclosure to social relations. In: Proceedings of the SIGCHI conference on Human factors in computing systems - CHI '05, p. 81. ACM Press, New York (2005). https://doi.org/10.1145/1054972.1054985

36. Cook, R.D., Weisberg, S.: Diagnostics for heteroscedasticity in regression. Biometrika 70(1), pp. 1–10 (1983). https://doi.org/10.1093/biomet/70.1.1

37. Cronbach, L.J.: Coefficient alpha and the internal structure of tests. Psychometrika 16(3), 297–334 (1951). https://doi.org/10.1007/BF02310555

38. Culnan, M.J., Armstrong, P.K.: Information privacy concerns, procedural fairness, and impersonal trust: an empirical investigation. Organ. Sci. 10(1), 1–115 (1999)

39. Cvrcek, D., Kumpost, M., Matyas, V., Danezis, G.: A study on the value of location privacy. In: Proceedings of the 5th ACM workshop on Privacy in Electronic Society, pp. 109–118 (2006). https://doi.org/10.1145/1179601.1179621

40. Danezis, G., Lewis, S., Anderson, R.: How much is location privacy worth? Fourth Workshop Econ. Inf. Secur. 78, 739–748 (2005). http://citeseerx.ist.psu.edu/viewdoc/download?doi=10.1.1.61.236&rep=rep1&type=pdf

41. Dardari, D., Closas, P., Djuric, P.M.: Indoor tracking: theory, methods, and technologies. IEEE Trans. Veh. Technol. 64(4), 1263–1278 (2015). https://doi.org/10.1109/TVT.2015.2403868

42. Davis, F.D.: Perceived usefulness, perceived ease of use, and user acceptance of information technology. MIS Q. 13(3), 319–340 (1989). https://doi.org/10.2307/249008

43. Debatin, B., Lovejoy, J.P., Horn, A.K., Hughes, B.N.: Facebook and online privacy: attitudes, behaviors, and unintended consequences. J. Comput.-Mediat. Commun. 15(1), 83–108 (2009). https://doi.org/10.1111/j.1083-6101.2009.01494.x

44. Deci, E.L., Ryan, R.M.: The "What" and "Why" of goal pursuits: human needs and the self-determination of behavior. Psychoanal. Inq. 11(4), 227–268 (2000)

45. Deutscher Bundestag: Basic Law for the Federal Republic of Germany in the revised version published in the Federal Law Gazette Part III, classification number 100-1, as last amended by Article 1 of the Act of 23 December 2014 (Federal Law Gazette I, p. 2438) (2014)

46. Dienlin, T., Trepte, S.: Is the privacy paradox a relic of the past? An in-depth analysis of privacy attitudes and privacy behaviors. Eur. J. Soc. Psychol. 45(3), 285–297 (2015). https://doi.org/10.1002/ejsp.2049

47. Dinev, T., Bellotto, M., Hart, P., Russo, V., Serra, I., Colautti, C.: Privacy calculus model in e-commerce—a study of Italy and the United States. Eur. J. Inf. Syst. 15(4), 389–402 (2006)

48. Dinev, T., Hart, P.: An extended privacy calculus model for e-commerce transactions. Inf. Sys. Res. 17(1), 61–80 (2006)

49. Dommeyer, C.J., Gross, B.L.: What consumers know and what they do: an investigation of consumer knowledge, awareness, and use of privacy protection strategies. J. Interact. Market. 17(2), 34–51 (2003). https://doi.org/10.1002/dir.10053

50. Duckham, M., Kulik, L.: A formal model of obfuscation and negotiation for location privacy. In: International Conference on Pervasive Computing. Springer, Berlin (2005)

51. Dwork, C.: Differential privacy. In: Lecture Notes in Computer Science (Including Subseries Lecture Notes in Artificial Intelligence and Lecture Notes in Bioinformatics) (2006). https://doi.org/10.1007/11787006-1
52. Engelbrecht-Wiggans, R., Katok, E.: E-sourcing in procurement: theory and behavior in reverse auctions with non-competitive contracts. Manag. Sci. 52(4), 581–596 (2005). https://doi.org/10.2139/ssrn.673562
53. European Court of Human Rights: Convention for the Protection of Human Rights and Fundamental Freedoms (1950). https://www.echr.coe.int/Documents/Convention_ENG.pdf
54. European Space Agency: Galileo: A Constellation of Navigation Satellites (2016). http://www.esa.int/Our_Activities/Navigation/Galileo/Galileo_a_constellation_of_navigation_satellites
55. Fabrigar, L.R., Wegener, D.T., MacCallum, R.C., Strahan, E.J.: Evaluating the use of exploratory factor analysis in psychological research. Psychol. Methods 4(3), 272 (1999). https://doi.org/10.1037/1082-989X.4.3.272
56. Felt, A.P., Chin, E., Hanna, S., Song, D., Wagner, D.: Android permissions demystified. In: Proceedings of the 18th ACM Conference on Computer and Communications Security - CCS '11, p. 627 (2011). http://dl.acm.org/citation.cfm?doid=2046707.2046779
57. Fischhoff, B., Slovic, P., Lichtenstein, S., Read, S., Combs, B.: How safe is safe enough? A psychometric study of attitudes towards technological risks and benefits. Policy Sci. 9(2), 127–152 (1978)
58. Fishbein, M., Ajzen, I.: Belief, attitude, intention and behaviour: an introduction to theory and research (1975). http://people.umass.edu/aizen/f&a1975.html
59. Fishbein, M., Ajzen, I.: Predicting and Changing Behavior: The Reasoned Action Approach. Predicting and Changing Behavior: The Reasoned Action Approach. Psychology Press, New York (2011)
60. Foucault, M.: Discipline and punish: the birth of the prison. Contemp. Sociol. (1978). https://doi.org/10.2307/2065008
61. Freudiger, J., Shokri, R., Hubaux, J.P.: Evaluating the privacy risk of location-based services. In: Lecture Notes in Computer Science (Including Subseries Lecture Notes in Artificial Intelligence and Lecture Notes in Bioinformatics). Lecture Notes in Computer Science, vol. 7035, pp. 31–46 (2012)
62. Goldfarb, A., Tucker, C.: Online display advertising: targeting and obtrusiveness. Market. Sci. 30(3), 413–415 (2011)
63. Golle, P., Partridge, K.: On the anonymity of home/work location pairs. In: Lecture Notes in Computer Science (Including Subseries Lecture Notes in Artificial Intelligence and Lecture Notes in Bioinformatics). Lecture Notes in Computer Science, vol. 5538, pp. 390–397 (2009)
64. Goodwin, L.D., Leech, N.L.: Understanding correlation: factors that affect the size of r. J. Exp. Educ. 74(3), 249–266 (2006). https://doi.org/10.3200/JEXE.74.3.249-266
65. Gosling, S.D., Rentfrow, P.J., Swann, W.B.: A very brief measure of the Big-Five personality domains. J. Res. Pers. 37(6), 504–528 (2003)
66. Grossklags, J., Hall, S., Acquisti, A.: When 25 Cents is too much: an experiment on willingness-to-sell and willingness-to-protect personal information. Information Security, pp. 7–8 (2007). http://citeseerx.ist.psu.edu/viewdoc/download?doi=10.1.1.137.696&rep=rep1&type=pdf
67. Grunewald, E.: FlashPoll: a location-aware polling application for iPhone: examining differences between iOS and android users in the context of mobile polling applications. Ph.D. Thesis, Chalmers University of Technology, Göteborg (2015)
68. Gruteser, M., Grunwald, D.: Anonymous usage of location-based services through spatial and temporal cloaking. In: Proceedings of the 1st International Conference on Mobile Systems, Applications and Services - MobiSys '03 (2003). https://doi.org/10.1145/1066116.1189037
69. Günther, O., Spiekermann, S.: RFID and the perception of control: the consumer's view. Commun. ACM 48(9), 73–76 (2005). https://doi.org/10.1145/1081992.1082023
70. Hair, J.F., Black, W.C., Babin, B.J., Anderson, R.E.: Multivariate Data Analysis. Pearson, New Jersey (2010). https://doi.org/10.1016/j.ijpharm.2011.02.019

71. Homburg, C., Giering, A.: Personal characteristics as moderators of the relationship between customer satisfaction and loyalty—an empirical analysis. Psychol. Market. **18**(1), 43–66 (2001). https://doi.org/10.1002/1520-6793(200101)18:1<43::AID-MAR3>3.0.CO;2-I

72. Hoofnagle, C.J., Urban, J.M.: Alan Westin's privacy homo economicus. Wake Forest Law Rev. **49**(2), 261–317 (2014)

73. Houghton, D.J., Joinson, A.N.: Privacy, social network sites, and social relations. J. Technol. Hum. Serv. **28**(1–2), 74–94 (2010). https://doi.org/10.1080/15228831003770775

74. Hoy, M., Milne, G.: Gender differences in privacy-related measures for young adult Facebook users. J. Interact. Advert. **10**(2), 28–45 (2010). http://www.tandfonline.com/doi/abs/10.1080/15252019.2010.10722168

75. Hu, L., Bentler, P.M.: Cutoff criteria for fit indexes in covariance structure analysis: conventional criteria versus new alternatives. Struct. Equ. Model. Multidiscip. J. **6**(1), 1–55 (1999). https://doi.org/10.1080/10705519909540118

76. Iachello, G., Hong, J.: End-user privacy in human-computer interaction. Found. Trends® Hum. Comput. Interact. **1**(1), 1 (2007). https://doi.org/10.1561/1100000004

77. Iachello, G., Smith, I., Consolvo, S., Abowd, G.D., Hughes, J., Howard, J., Potter, F., Scott, J., Sohn, T., Hightower, J., Lamarca, A.: Control , deception , and communication: evaluating the deployment of a location-enhanced messaging service. In: UbiComp 2005: Ubiquitous Computing, pp. 213–231 (2005). https://doi.org/10.1007/11551201-13

78. Ignatius, E., Kokkonen, M.: Factors contributing to verbal self-disclosure. Nordic Psychol. **59**(4), 362–391 (2007). https://doi.org/10.1027/1901-2276.59.4.362

79. Jensen, C., Potts, C.: Privacy policies as decision-making tools. In: Proceedings of the 2004 Conference on Human Factors in Computing Systems - CHI '04, vol. 6, pp. 471–478 (2004). https://doi.org/10.1145/985692.985752

80. Jensen, C., Potts, C., Jensen, C.: Privacy practices of Internet users: self-reports versus observed behavior. Int. J. Hum. Comput. Stud. **63**(1–2), 203–227 (2005)

81. John, L.K., Acquisti, A., Loewenstein, G.: Strangers on a plane: context-dependent willingness to divulge sensitive information. J. Consum. Res. **37**(5), 858–873 (2011). http://www.scopus.com/inward/record.url?eid=2-s2.0-78751545877&partnerID=tZOtx3y1

82. Joinson, A.N., Paine, C., Buchanan, T., Reips, U.D.: Measuring self-disclosure online: blurring and non-response to sensitive items in web-based surveys. Comput. Hum. Behav. **24**(5), 2158–2171 (2008)

83. Junglas, I., Spitzmüller, C.: Personality traits and privacy perceptions: an empirical study in the context of location-based services. In: International Conference on Mobile Business, ICMB 2006 (2006). https://doi.org/10.1109/ICMB.2006.40

84. Kachore, V.A., Lakshmi, J., Nandy, S.K.: Location obfuscation for location data privacy. In: Proceedings of 2015 IEEE World Congress on Services, SERVICES 2015 (2015). https://doi.org/10.1109/SERVICES.2015.39

85. Kahneman, D., Knetsch, J.L., Thaler, R.H.: Experimental tests of the endowment effect and the coase theorem. J. Polit. Econ. **98**(6), 1325–1348 (1990)

86. Kahneman, D., Slovic, P., Tversky, A.: Judgment under uncertainty: heuristics and biases. Science **185**(4157), 1124–1131 (1974). https://doi.org/10.1126/science.185.4157.1124

87. Kahneman, D., Tversky, A.: Prospect theory: an analysis of decision under risk. Econometrica **47**(2), 263–292 (1979). https://doi.org/10.2307/1914185

88. Kelley, P.G., Benisch, M., Cranor, L.F., Sadeh, N.: When are users comfortable sharing locations with advertisers? In: Proceedings of the SIGCHI Conference on Human Factors in Computing Systems, pp. 2449–2452 (2011). https://doi.org/10.1145/1978942.1979299

89. Kelley, P.G., Bresee, J., Cranor, L.F., Reeder, R.W.: A "nutrition label" for privacy. In: Proceedings of the 5th Symposium on Usable Privacy and Security - SOUPS (2009). http://portal.acm.org/citation.cfm?doid=1572532.1572538

90. Kelley, P.G., Consolvo, S., Cranor, L.F., Jung, J., Sadeh, N., Wetherall, D.: A conundrum of permissions: Installing applications on an android smartphone. In: Lecture Notes in Computer Science (Including Subseries Lecture Notes in Artificial Intelligence and Lecture Notes in Bioinformatics). Lecture Notes in Computer Science, vol. 7398, pp. 68–79 (2012)

91. Kolenikov, S., Bollen, K.A.: Testing negative error variances: is a Heywood case a symptom of misspecification? Sociol. Methods Res. **41**(1), 124–167 (2012). https://doi.org/10.1177/0049124112442138

92. Krasnova, H., Spiekermann, S., Koroleva, K., Hildebrand, T.: Online social networks: why we disclose. J. Inf. Technol. **25**(2), 109–125 (2010)

93. Kraus, L., Hirsch, T., Wechsung, I., Poikela, M., Möller, S.: Poster: towards an instrument to measure everyday privacy and security knowledge. In: Symposium on Usable Privacy and Security (SOUPS) (2014)

94. Kraus, L., Wechsung, I., Moller, S.: Using statistical information to communicate android permission risks to users. In: 4th Workshop on Socio-Technical Aspects in Security and Trust (STAST), pp. 48–55. Institute of Electrical and Electronics Engineers, Piscataway (2014)

95. Kumaraguru, P., Cranor, L.: Privacy indexes: a survey of westin's studies. Science Tech. rep. (December), 1–22 (2005). http://www.casos.cs.cmu.edu/publications/papers/CMU-ISRI-05-138.pdf%5Cnhttp://repository.cmu.edu/isr/856/

96. Küpper, A.: Location-Based Services: Fundamentals and Operation. Wiley, London (2005). https://doi.org/10.1002/0470092335

97. Lederer, S., Mankoff, J., Dey, A.K.: Who wants to know what when? Privacy preference determinants in ubiquitous computing. In: CHI'03 Extended Abstracts on Human Factors in Computing Systems, p. 724 (2003). https://doi.org/10.1145/765891.765952.

98. Leszczynski, A.: Spatial big data and anxieties of control. Environ. Plan. D: Soc. Space **33**(6), 965–984 (2015)

99. Lin, J., Sadeh, N., Amini, S., Lindqvist, J., Hong, J.I., Zhang, J.: Expectation and purpose: understanding users' mental models of mobile app privacy through crowdsourcing. In: Proceedings of the 2012 ACM Conference on Ubiquitous Computing - UbiComp '12 (2012). https://doi.org/10.1145/2370216.2370290

100. Lindqvist, J., Cranshaw, J., Wiese, J., Hong, J., Zimmerman, J.: I'm the mayor of my house. In: Proceedings of the 2011 Annual Conference on Human Factors in Computing Systems CHI 11, p. 2409 (2011). https://doi.org/10.1145/1978942.1979295

101. Lu, Y., Tan, B., Hui, K.L.: Inducing customers to disclose personal information to internet businesses with social adjustment benefits. In: Proceedings of the International Conference on Information Systems, ICIS (2004)

102. Malhotra, N.K., Kim, S.S., Agarwal, J.: Internet users' information privacy concerns (IUIPC): the construct, the scale, and a causal model. Inf. Sys. Res. **15**(4), 336–355 (2004). https://doi.org/10.1287/isre.1040.0032

103. Mayer, R.C., Davis, J.H., Schoorman, F.D.: An integrative model of organizational trust. Acad. Manag. Rev. **20**(3), 709–734 (1995). https://doi.org/10.5465/AMR.1995.9508080335

104. McDonald, A., Cranor, L.: Beliefs and behaviors: internet users' understanding of behavioral advertising. In: Telecommunications Policy Research Conference (2010). https://doi.org/10.2139/ssrn.1989092

105. McDonald, A.M., Cranor, L.F.: Americans' attitudes about internet behavioral advertising practices. In: Proceedings of the 9th Annual ACM Workshop on Privacy in the Electronic Society - WPES '10 (2010). https://doi.org/10.1145/1866919.1866929

106. Miao, F., He, Y., Liu, J., Li, Y., Ayoola, I.: Identifying typical physical activity on smartphone with varying positions and orientations. BioMed. Eng. Online **14**(1), 32 (2015). https://doi.org/10.1186/s12938-015-0026-4

107. Monahan, T., Fisher, J.A.: Benefits of 'observer effects': lessons from the field. Qual. Res. **10**(3), 357–376. (2010). https://doi.org/10.1177/1468794110362874

108. Mongin, P.: Expected utility theory. In: Handbook of Economic Methodology (1998). https://doi.org/10.2139/ssrn.1033982

109. de Montjoye, Y.A., Hidalgo, C.A., Verleysen, M., Blondel, V.D.: Unique in the crowd: the privacy bounds of human mobility. Sci. Rep. **3**, 1376 (2013). http://www.pubmedcentral.nih.gov/articlerender.fcgi?artid=3607247&tool=pmcentrez&rendertype=abstract

110. Moosbrugger, H., Kelava, A.: Testtheorie und Fragebogen- konstruktion (2012). https://doi.org/10.1007/978-3-642-20072-4-2

111. Morton, A.: Measuring inherent privacy concern and desire for privacy—a pilot survey study of an instrument to measure dispositional privacy concern. In: International Conference on Social Computing (SocialCom) (2013)

112. Murphy, W., Hereman, W.: Determination of a position in three dimensions using trilateration and approximate distances (1995). https://doi.org/10.1017/CBO9781107415324.004

113. Naderi, B.: Motivation of Workers on Microtask Crowdsourcing Platforms. Springer, Berlin (2017)

114. Naderi, B., Polzehl, T., Wechsung, I., Köster, F., Möller, S.: Effect of trapping questions on the reliability of speech quality judgments in a crowdsourcing paradigm. In: Proceedings of the Annual Conference of the International Speech Communication Association, INTER-SPEECH (2015)

115. Nederhof, A.J.: Methods of coping with social desirability bias: a review. Eur. J. Soc. Psychol. **15**(3), 263–280 (1985). https://doi.org/10.1002/ejsp.2420150303

116. Nissenbaum, H.: Privacy as contextual integrity. Wash. Law Rev. **79**(1), 101–139 (2004). https://doi.org/10.1109/SP.2006.32

117. Norberg, P.A., Horne, D.R., Horne, D.A.: The privacy paradox: personal information disclosure intentions versus behaviors. J. Consum. Aff. **41**(1), 100–126 (2007)

118. Nunnally, J., Bernstein, I.: Psychometric Theory, 3rd edn. McGraw-Hill, New York (1994)

119. Olson, J.S., Arbor, A., Grudin, J., Horvitz, E.: Toward understanding preferences for sharing and privacy. In: CHI '05 Extended Abstracts on Human Factors in Computing Systems, pp. 1985–1988 (2004). ftp://ftp.research.microsoft.com/pub/tr/TR-2004-138.pdf

120. Oulasvirta, A., Suomalainen, T., Hamari, J., Lampinen, A., Karvonen, K.: Transparency of intentions decreases privacy concerns in ubiquitous surveillance. Cyberpsychol. Behav. Soc. Netw. **17**(10), 633–638 (2014)

121. Palen, L., Dourish, P.: Unpacking "privacy" for a networked world. In: Proceedings of the conference on Human factors in computing systems - CHI '03, vol. 5, p. 129 (2003). https://doi.org/10.1145/642633.642635

122. Palmer, M.: Data is the New Oil (2006)

123. Park, Y.J., Mo Jang, S.: Understanding privacy knowledge and skill in mobile communication. Comput. Hum. Behav. **38**, 296–303 (2014). https://doi.org/10.1016/j.chb.2014.05.041

124. Patil, S., Norcie, G., Kapadia, A., Lee, A.J.: Reasons, rewards, regrets: privacy considerations in location sharing as an interactive practice. In: Proceedings of the Eighth Symposium on Usable Privacy and Security (SOUPS '12), pp. 1–15. ACM, New York (2012)

125. Patil, S., Schlegel, R., Kapadia, A., Lee, A.J.: Reflection or action? How feedback and control affect location sharing decisions. In: Proceedings of the SIGCHI Conference on Human Factors in Computing Systems (CHI '14), pp. 101–110. ACM, New York (2014)

126. Pedersen, D.M.: Model for types of privacy by privacy functions. J. Environ. Psychol. **19**(4), 397–405 (1999). https://doi.org/10.1006/jevp.1999.0140

127. Petronio, S.: Boundaries of Privacy: Dialectics of disclosure. State University of New York Press, New York (2002)

128. Petronio, S., Durham, W.T.: Communication privacy management theory (2008). http://ovidsp.ovid.com/ovidweb.cgi?T=JS&PAGE=reference&D=psyc6&NEWS=N&AN=2008-05087-023

129. Pew Research Center: Mobile Fact Sheet (2018)

130. Phelan, C., Lampe, C., Resnick, P.: It's creepy, but it doesn't bother me. In: CHI '16 Proceedings of the 2016 CHI Conference on Human Factors in Computing Systems, pp. 5240–5251. New York (2016). https://doi.org/10.1145/2858036.2858381

131. Pierson, C.T.: Data breaches highlight the importance of privacy. Financ. Exec. **25**(2), 62–64 (2009)

132. Pingley, A., Zhang, N., Fu, X., Choi, H.A., Subramaniam, S., Zhao, W.: Protection of query privacy for continuous location based services. In: Proceedings - IEEE INFOCOM (2011). https://doi.org/10.1109/INFCOM.2011.5934968

133. Poikela, M., Kaiser, F.: "It is a topic that confuses me"—privacy perceptions in usage of location-based applications. In: European Workshop on Usable Security. Internet Society, Geneva (2016)

134. Poikela, M., Schmidt, R., Wechsung, I., Möller, S.: Locate! When do users disclose location? In: Workshop on Privacy Personas and Segmentation (PPS) at the Tenth Symposium On Usable Privacy and Security (SOUPS). USENIX Association, Menlo Park (2014)

135. Poikela, M., Schmidt, R., Wechsung, I., Möller, S.: "About your smartphone usage"—privacy in location-based mobile participation. In: IEEE International Symposium on Technology and Society (ISTAS 2015). IEEE, Dublin (2015)

136. Poikela, M., Schmidt, R., Wechsung, I., Möller, S.: "About your smartphone usage"–privacy in location-based mobile participation. In: Proceedings of the International Symposium on Technology and Society, vol. 2016-March (2016). https://doi.org/10.1109/ISTAS.2015.7439421

137. Poikela, M., Toch, E.: Understanding the valuation of location privacy: a crowdsourcing-based approach. In: Proceedings of the 50th Annual Hawaii International Conference on System Sciences (2017)

138. Poikela, M., Wechsung, I., Möller, S.: Location-based applications-benefits, risks, and concerns as usage predictors. In: 2nd Annual Privacy Personas and Segmentation (PPS) Workshop at the Eleventh Symposium on Usable Privacy and Security (SOUPS). Ottawa (2015)

139. Popovic, M., Milne, D., Barrett, P.: The scale of perceived interpersonal closeness (PICS). Clin. Psychol. Psychother. 10(5), 286–301 (2003). https://doi.org/10.1002/cpp.375

140. Preibusch, S.: Privacy behaviors after Snowden. Commun. ACM 58(5), 48–55 (2015). https://doi.org/10.1145/2663341

141. Rose, E.: Data users versus data subjects: are consumers willing to pay for property rights to personal information? In: Proceedings of the 38th Annual Hawaii International Conference on System Sciences (2005)

142. Ryan, R.M., Kuhl, J., Deci, E.L.: Nature and autonomy: an organizational view of social and neurobiological aspects of self-regulation in behavior and development. Dev. Psychobiol. 9(4), 701–728 (1997)

143. Sadeh, N., Hong, J., Cranor, L., Fette, I., Kelley, P., Prabaker, M., Rao, J.: Understanding and capturing people's privacy policies in a mobile social networking application. In: Personal and Ubiquitous Computing, vol. 13, pp. 401–412 (2009)

144. Schaub, F., Balebako, R., Durity, A.L., Cranor, L.F.: A design space for effective privacy notices. In: Eleventh Symposium On Usable Privacy and Security (SOUPS 2015) (2015)

145. Schwarz, N.: Self-reports: How the questions shape the answers. Am. Psychol. 54(2), 93 (1999). https://doi.org/10.1037/0003-066X.54.2.93

146. Sheehan, K.B.: An investigation of gender differences in on-line privacy concerns and resultant behaviors. J. Interact. Market. 13(4), 24–38 (1999)

147. Sheehan, K.B.: Toward a typology of internet users and online privacy concerns. Inf. Soc. 18(1), 21–32 (2002)

148. Shklovski, I., Mainwaring, S.D., Skúladóttir, H.H., Borgthorsson, H.: Leakiness and creepiness in app space. In: Proceedings of the 32nd Annual ACM Conference on Human Factors in Computing Systems - CHI '14 (2014). https://doi.org/10.1145/2556288.2557421

149. Smith, H.J., Dinev, T., Xu, H.: Theory and review information privacy research: an interdisciplinary review. MIS Q. Inf. Privacy Res. 35(4), 989–1015 (2011)

150. Smith, H.J., Milberg, S.J., Burke, S.J.: Information privacy: measuring individuals' concerns about organizational practices. MIS Q. 20(2), 167–196 (1996). https://doi.org/10.2307/249477

151. Solove, D.J.: 'I've got nothing to hide' and other misunderstandings of privacy. San Diego Law Rev. 44, 1–23 (2007). https://doi.org/10.2139/ssrn.998565

152. Spiekermann, S., Cranor, L.F.: Engineering privacy. IEEE Trans. Softw. Eng. 35(1), 67–82 (2009)

153. Spiekermann, S., Grossklags, J., Berendt, B.: E-privacy in 2nd generation E-commerce: privacy preferences versus actual behavior. In: EC '01 Third ACM Conference on Electronic Commerce, pp. 38–47. Humboldt University Berlin, ACM, New York (2001)
154. Statista: Anteil der Smartphone-Nutzer in Deutschland in den Jahren 2012 bis 2017 (2018)
155. Statista: Number of apps available in leading app stores as of 1st quarter 2018. Tech. rep. (2018). https://www.statista.com/statistics/276623/number-of-apps-available-in-leading-app-stores/
156. Stutzman, F., Kramer-Duffield, J.: Friends only: examining a privacy-enhancing behavior in Facebook. In: Proceedings of the SIGCHI Conference on Human Factors in Computing Systems, pp. 1553–1562 (2010). http://dl.acm.org/citation.cfm?id=1753559
157. Sweeney, L.: K-anonymity: a model for protecting privacy. Int. J. Uncertainty Fuzziness Knowledge Based Syst. **10**(5), 557–570 (2002). https://doi.org/10.1142/S0218488502001648
158. Tang, K., Lin, J., Hong, J.: Rethinking location sharing: exploring the implications of social-driven vs. purpose-driven location sharing. In: Proceedings of the 12th ACM International Conference on Ubiquitous Computing - Ubicomp '10, vol. 12(4–5), 85–94 (2010). https://doi.org/10.1145/1864349.1864363
159. Thiesse, F.: RFID, privacy and the perception of risk: A strategic framework. J. Strateg. Inf. Syst. **16**(2), 214–232 (2007). https://doi.org/10.1016/j.jsis.2007.05.006
160. Tikkinen-Piri, C., Rohunen, A., Markkula, J.: EU general data protection regulation: changes and implications for personal data collecting companies. Comput. Law Secur. Rev. **34**(1), 134–153 (2018). https://doi.org/10.1016/j.clsr.2017.05.015
161. Toch, E., Cranshaw, J., Drielsma, P.H., Tsai, J.Y., Kelley, P.G., Springfield, J., Cranor, L., Hong, J., Sadeh, N.: Empirical models of privacy in location sharing. In: Proceedings of the 12th ACM international conference on Ubiquitous computing - Ubicomp '10, p. 129 (2010). https://doi.org/10.1145/1864349.1864364
162. Tsai, J.Y., Egelman, S., Cranor, L., Acquisti, A.: The effect of online privacy information on purchasing behavior: an experimental study. Inf. Syst. Res. **22**(2), 254–268 (2011). https://doi.org/10.1287/isre.1090.0260
163. Tsai, J.Y., Kelley, P.G., Cranor, L.F., Sadeh, N.: Location-sharing technologies : privacy risks and controls. J. Law Policy Inf. Soc. **6**(2), 119–151 (2010)
164. Turow, J., Feldman, L., Meltzer, K.: Open to Exploitation: America's Shoppers Online and Offline. In: Annenberg Public Policy Center, p. 10 (2005). http://repository.upenn.edu/cgi/viewcontent.cgi?article=1035&context=asc_papers
165. Utz, S., Kramer, N.C.: The privacy paradox on social network sites revisited: the role of individual characteristics and group norms. J. Psychosoc. Res. Cyberspace **3**(2), 1–12 (2009)
166. Venkatanathan, J., Lin, J., Benisch, M., Ferreira, D., Karapanos, E., Kostakos, V., Sadeh, N., Toch, E.: Who, when, where: obfuscation preferences in location-sharing applications (2011). https://doi.org/papers3://publication/uuid/50230E56-81A4-4FFD-AE05-5463CEE31F3B
167. Venkatesh, V., Morris, M.G., Davis, G.B., Davis, F.D.: User acceptance of information technology: toward a unified view. MIS Q. **27**(3), 425–478 (2003). https://doi.org/10.2307/30036540
168. Wang, E.S.T., Lin, R.L.: Perceived quality factors of location-based apps on trust, perceived privacy risk, and continuous usage intention. Behav. Inf. Technol. **36**(1), 2–10 (2017). https://doi.org/10.1080/0144929X.2016.1143033
169. Wang, Y., Leon, P.G., Acquisti, A., Cranor, L.F., Forget, A., Sadeh, N.: A field trial of privacy nudges for Facebook. In: Proceedings of the 32nd annual ACM Conference on Human Factors in Computing Systems - CHI '14, pp. 2367–2376 (2014). http://dl.acm.org/citation.cfm?id=2556288.2557413
170. Wang, Y., Norcie, G., Komanduri, S., Acquisti, A., Leon, P.G., Cranor, L.F.: "I regretted the minute I pressed share". In: Proceedings of the Seventh Symposium on Usable Privacy and Security - SOUPS '11 (2011). https://doi.org/10.1145/2078827.2078841
171. Wang, Z., Yang, Z., Dong, T.: A review of wearable technologies for elderly care that can accurately track indoor position, recognize physical activities and monitor vital signs in real time. Sensors, **17**(2), 341 (2017). https://doi.org/10.3390/s17020341

172. Wechsung, I.: An Evaluation Framework for Multimodal Interaction. T-Labs Series in Telecommunication Services, vol. 10, pp. 973–978. Springer, Cham (2014)

173. Wei, W., Xu, F., Li, Q.: MobiShare: flexible privacy-preserving location sharing in mobile online social networks. In: Proceedings - IEEE INFOCOM (2012). https://doi.org/10.1109/INFCOM.2012.6195664

174. Westin, A.F.: Social and political dimensions of privacy. J. Soc. Issues **59**(2), 431–453 (2003)

175. Whitman, J.Q.: The two western cultures of privacy: dignity versus liberty. Yale LJ **113**, 1151 (2004). https://doi.org/10.2139/ssrn.476041

176. Xu, H., Gupta, S.: The effects of privacy concerns and personal innovativeness on potential and experienced customers' adoption of location-based services. Electron. Mark. **19**(2–3), 137–149 (2009). https://doi.org/10.1007/s12525-009-0012-4

177. Xu, H., Gupta, S., Rosson, M.B., Carroll, J.M.: Measuring mobile users' concerns for information privacy. ICIS (Ftc 2009), pp. 1–16 (2012). http://faculty.ist.psu.edu/xu/papers/Xu_etal_ICIS_2012a.pdf

178. Xu, H., Luo, X., Carroll, J.M., Rosson, M.B.: The personalization privacy paradox: An exploratory study of decision making process for location-aware marketing. Decis. Support Syst. **51**(1), 42–52 (2011)

179. Xu, H., Teo, H.h., Tan, B.C.Y.: Predicting the adoption of location-based services: the role of trust and perceived privacy risk. In: Proceedings of the 26th International Conference on Information Systems (ICIS 2005), pp. 897–910. Las Vegas (Beinat 2001) (2005)

180. Xu, H., Teo, H.H., Tan, B.C.Y., Agarwal, R.: Effects of individual self-protection, industry self-regulation, and government regulation on privacy concerns: a study of location-based services. Inf. Syst. Res. **23**(4), 1342–1363 (2012)

181. Young, A.L., Quan-Haase, A.: Privacy protection strategies on Facebook: the internet privacy paradox revisited. Inf. Commun. Soc. **16**(4), 479–500 (2013). https://doi.org/10.1080/1369118X.2013.777757

182. Zhang, L., Tiwana, B., Qian, Z., Wang, Z., Dick, R.P., Mao, Z.M., Yang, L.: Accurate online power estimation and automatic battery behavior based power model generation for smartphones. In: Proceedings of the Eighth IEEE/ACM/IFIP International Conference on Hardware/Software Codesign and System Synthesis - CODES/ISSS '10 (2010). https://doi.org/10.1145/1878961.1878982

183. Zhou, T.: The impact of privacy concern on user adoption of location-based services. Ind. Manag. Data Syst. **111**(2), 212–226 (2011). https://doi.org/10.1108/02635571111115146

184. Zhou, T.: Examining location-based services usage from the perspectives of unified theory of acceptance and use of technology and privacy risk. J. Electron. Commer. Res. **13**(2), 135–144 (2012)

Index

© Springer Nature Switzerland AG 2020
M. E. Poikela, *Perceived Privacy in Location-Based Mobile System*,
T-Labs Series in Telecommunication Services,
https://doi.org/10.1007/978-3-030-34171-8

Printed in the United States
By Bookmasters